Necessary Inventions

NECESSARY INVENTIONS

Roger Bacon, the Middle Ages, and the Making of Modern Science

E. R. Truitt

PENN

UNIVERSITY OF PENNSYLVANIA PRESS

PHILADELPHIA

Published by
University of Pennsylvania Press
Philadelphia, Pennsylvania 19104–4112 USA
www.pennpress.org

EU Authorized Representative: Easy Access System Europe—Mustamäe tee 50, 10621 Tallinn, Estonia, gpsr.requests@easproject.com.

Printed in the United States of America on acid-free paper
10 9 8 7 6 5 4 3 2 1

A Cataloging-in-Publication record is available from the Library of Congress
Hardcover ISBN: 978-1-5128-2988-4
eBook ISBN: 978-1-5128-2989-1

For M. Sean Grady and Jang Won Yoon, with thanks

Contents

Introduction

Invention and Innovation

> Now, although Nature is powerful and marvelous, yet art, using Nature as an instrument, is more powerful than natural strength, as we see in many cases.
>
> —Roger Bacon, *Letter on the Hidden Powers of Art and Nature, and the Invalidity of Magic*, ca. 1270s

> The present generation finds itself to be the heir of a vast patrimony of science; and it must needs concern us to know the steps by which these possessions were acquired. . . . Our species, from the time of its creation, has been travelling onwards in pursuit of truth; and now that we have reached a lofty and commanding position, with the broad light of day around us, it must be grateful to look back on the line of our past progress.
>
> —William Whewell, *History of the Inductive Sciences*, 1837

> Only the civilizations descended from Hellenic Greece have possessed more than the most rudimentary science. The bulk of scientific knowledge is a product of Europe in the last four centuries.
>
> —Thomas Kuhn, *The Structure of Scientific Revolutions*, 1967

In his *Letter on the Hidden Powers of Art and Nature, and the Invalidity of Magic*, English Scholastic philosopher and Franciscan friar Roger Bacon (ca. 1214–92) invoked an enticing future in which horseless chariots would move at "unimaginable speed," submarines and diving bells would enable undersea exploration, flying machines would allow aerial reconnaissance, alchemical and pharmaceutical interventions would restore health and extend human lifespan, and mirrors and lenses would set fire to entire armies or produce optical illusions to evoke delight or "infinite terrors."[1] In this short treatise Bacon asserted that natural knowledge, encompassing theory and sensory knowledge, combined with art, or technology, could surpass magic in its effects, foster astonishing inventions and new knowledge, and recover lost wisdom.[2] A subtle understanding of nature and the inventions that harnessed nature's powers could also be turned to political and religious advantage, promising a future of Christian political and military domination and the spread of Christian values.

These speculative technologies and possible futures represent the promise of *scientia experimentalis*, Bacon's term for his method of natural knowledge-acquisition and knowledge-production.[3] In conceiving this method, he was influenced by the Aristotelian natural philosophy that he studied and eventually taught at the University of Paris, and also by contemporary intellectual currents focused on the sensorium and observation, texts on statecraft and history, and work by earlier scholars working in Arabic. *Scientia experimentalis* relied on both syllogistic reasoning (*argumentum*) and sensory knowledge (*experimentum*) to attain accurate knowledge about the natural world. The purpose of this knowledge was, according to Bacon, to help the individual become a more perfect Christian and help Latin Christendom—or a Christian nation like England—become more powerful. The devices Bacon wrote about are a product of *scientia experimentalis* as well as necessary to further inquiry. Such inquiry had several connected and mutually reinforcing aims: to know more, and more completely, about nature and to use that knowledge for personal edification and political empowerment.

The possibilities for enlightenment and domination afforded by *scientia experimentalis* and the technologies it might realize kindled the imagination of many. *Hidden Powers* and other texts by Bacon on *scientia experimentalis* circulated widely throughout the Latin West in the centuries after his death. These included works on optics (*perspectiva*) and alchemy, along with spurious texts attributed to Bacon but not authored by him. Philosophers, scholars, and antiquarians in the sixteenth and seventeenth centuries constructed Bacon's ingenuity and anticipation, as well as his supposedly subversive or illicit practices, through productive, partial readings of his work, especially relating to *scientia experimentalis*. Moreover, by the end of the seventeenth century Bacon was reimagined in a specifically English context, despite his extensive ties to Paris and, to a lesser extent, Rome. Revenant versions of "Roger Bacon" have appeared in the centuries since his death in 1292: among them, avid experimentalist, reluctant teacher, conjuring friar, and alchemist. This interest on the part of later writers continued after the seventeenth century and eventually configured Bacon's reputation for inventiveness as wholly positive.

By the nineteenth century, as William Whewell (1794–1866) and others defined science as a specific method of natural knowledge-acquisition with an intellectual history that began with ancient Greek philosophy and was associated with mental ability, Roger Bacon appeared within that history as a brilliant visionary.[4] After Whewell, Roger Bacon's reputation began to rival that of the so-called father of empiricism and of English science, Francis Bacon (1561–1626).[5] In the nineteenth and twentieth centuries Roger Bacon was often presented as a man committed to rational inquiry and the refutation of error and superstition, a Leonardo-like figure ahead of his time who forecast (or, according to some, invented) later technologies, and an experimentalist whose theory of knowledge acquisition anticipated, by centuries, the scientific method of inductive reasoning via experiment.[6] Although there is some truth in this representation—Bacon focused on questions of knowledge-production and the eradication of error, and prioritized knowledge through experiment—the contours of myth and the gaps between myth and reality reveal later preoccupations. Whewell, in his *History of the Inductive Sciences*, first published in 1837, set the terms of subsequent scholarship on Bacon around precocity and influence, marveling that "Roger Bacon's works are not only so far beyond his age in the knowledge which they contain, but so different from the temper of the times, in his assertion of the supremacy of experiment, and in his contemplation of the future progress of knowledge, that it is difficult to conceive how such a character could then exist."[7]

But why is it difficult? What ideas or assumptions about the thirteenth century give rise to this difficulty? Many of Bacon's ideas about the importance of observation, testing, and empiricism in the pursuit of natural knowledge resemble those of later thinkers, including Francis Bacon and William Whewell. So why articulate that resemblance in terms of anachronism and precocity, rather than likeness or reverberation? What does it mean to say of Bacon, or of any long-dead figure, *he is like us* rather than *we are like him?*

Historical Periodization and Temporal Dislocation

The framing of *he is like us* in historical context articulates a relationship between past and present that assimilates an element of the past into the present. This relationship is an inversion of the dynamic Johannes Fabian outlined in his 1983 work *Time and the Other.* In cultural anthropology, the ethnographer and the ethnographic subject inhabit the same time, as coevals, while the ethnographer is "in the field." However, at the desk, researchers deny coeval status to the ethnographic subject by placing the "primitive" or "savage" into an unchanging and eternal present, outside of the ethnographer's experience of time.[8] Fabian noted that this temporal differentiation is political and used to universalize Western progress and justify Western colonialism. Postcolonial theorists and scholars of the medieval period have used Fabian's framework to consider the relationship between postcolonialism and medievalism; I enjoin my efforts in the history of science to theirs.[9] Using Fabian's framework to think about the relationship between the historian and the historical subject, I argue that the kind of exceptionalism accorded to Bacon, articulated here by Whewell, also participates in the same relationship to time for similar purposes. However, where the ethnographer and ethnographic subject occupy the same time during the period of research, the historian and the historical subject begin in different time periods. Saying that Roger Bacon or any other long-dead person is "like us" articulates a coeval relationship between the historian and their historical subject. This temporal dislocation is likewise not neutral; it permits the historian to substitute their own context—their present—for that of their subject, ignoring what gives any historical fact, or text, or action its meaning: context.

Although I use Fabian's framework, I also critique the periodization on which his framework is predicated. Fabian traced the denial of coevality to the secularization of Christian time, which was linear and oriented toward salvation, and the subsequent shift, in Christian Europe, to seeing travel through space as travel back in time. In his account, traveling beyond the capital cities of Europe in the nineteenth and early twentieth centuries was a way to visit the past, because to see how rural or foreign people live "now" is to see how people—even the traveler's own ancestors—lived long ago. They live as they have always lived, in an eternal present just like the anthropological "primitive." Yet Fabian asserted that this secularization of time, which made it possible to view travel through space *as* travel backward through time, was a response to the "rediscovery" of science from the Greeks in the Renaissance, implicitly characterizing the long interval between those two epochs as a period that is neither modern nor leading to modernity.[10] The periodization on which Fabian relied also has a history; there is nothing natural about it. "Long ago and far away" is not a meaningless phrase to introduce a fairy tale; it is a displacement mechanism for collapsing the chronologically distant past and geographically or conceptually distant present into the same category of "other."

The idea that the period called the Middle Ages—the millennium between the political devolution of the western half of the Roman Empire in the fifth century and the fifteenth century—was one of intellectual and cultural decline has its own history. Furthermore, this period—the chronological boundaries, geographic contours, and linguistic families—has a complex history of definition and redefinition in historiography, which began during that millennium itself and took hold in what we now call the early modern period. These related ideas—that the period between the two that are now called classical antiquity and the early modern period constitutes a historical era and that this era is marked by widespread intellectual, political, and social stagnation and regression—are central to the historiography of science and the narrative of scientific progress. A detailed, expansive reading of the historiography of Bacon since the fourteenth century reveals how a definition of modern science as objective, rational, authoritative, and methodical relies on "the Middle Ages": a fantastical period characterized as credulous, superstitious, dogmatic, and erratic. The imagined thousand-year historical era called "the Middle Ages" renders possible the stark contrast with what comes after it—modernity. Roger Bacon, characterized as a modern mind trapped in an unmodern age, appears as a prophetic harbinger of progress in a stagnant time.

Whewell's narrative about scientific progress and the historical periodization that narrative supported proved enormously influential, echoed by Kuhn, Fabian, and countless others. Bacon's exceptionalism proves the rule that the Middle Ages were intellectually "stationary" and Bacon's ideas, taken as an example of scientific thinking, make possible a claim that modern science could only emerge in a European Christian context.[11] Excavating different versions of Roger Bacon framed after his death and exhibiting his establishment as a significant figure in the history of science exposes the production of a "great man" narrative out of a mix of contingencies, purposeful interventions, partial translations, and misattributions. This narrative emerged over time via selective reading, and in the service of European—mostly British—Christian nationalism. In the historiography of science through the twentieth century, Bacon was generally understood in the terms set by Whewell. His beliefs in the coming apocalypse, in knowledge through divine revelation, and in the role of natural knowledge to serve Christian morals have largely been omitted from the picture of "Roger Bacon, proto-scientist and secular experimentalist."[12]

I explore what unarticulated assumptions shape and uphold the notion that Bacon (or anyone) could be different from and ahead of his time and explain the consequences of those assumptions for the history of science. A close examination of Bacon's writings about *scientia experimentalis* and allied subjects, alongside his intellectual influences, from courtly histories to Arabic philosophy, brings into focus a picture of Bacon in which he is exemplary of his milieu, rather than exceptional for his time. This approach both explores the limitations of biography, especially intellectual biography, and offers the potential to re-script the genre, especially given a wider lens and longer timespan. By complicating and broadening the picture of Roger Bacon and of medieval science, I argue for a historiography of science that views periodization with some skepticism and frees the history of science from the tether of teleology.

A critical reexamination of Bacon's posthumous reputations and how they were created alongside his writings and those of his contemporaries and influences offers a case study in how the "Great Man" narrative, centered on an inventive or otherwise visionary man, operates to support historical periodization. This periodization, emergent in the fifteenth century and institutionalized in the nineteenth century, of ancient-medieval-modern, is itself a construct that rests on and further supports the ideology of Christian supremacy. One way in which the medieval period is perhaps surprisingly foundational to the historiography of science is in the importance of medieval Christian theological doctrine to ideas about how to divide the past into different eras. These ideas, which concern the relationship of the past to the present and future, are central to the traditional narrative of the emergence of science. The terms of scholarship on Bacon—precocity, influence, and anticipation—were established in the nineteenth century alongside the establishment of modern science as a set of practices and objectives, the codification of "the Middle Ages" as a historical period defined in opposition to modernity, and widespread expression of medievalism in the arts. These terms affirm later values and derive from an impulse to prospect the past to understand the present and to advance a progressive view of knowledge. This way of understanding history, and using it to articulate a relationship between past, present, and future, has been identified as a shift in the "regime of historicity" that took place between 1750 and 1850, crystallizing after the French Revolution and with the professionalization of history as a subject of study.[13] Yet, as I demonstrate, this relationship between past, present, and future emerged from Christian theology during late antiquity and became firmly entrenched by the thirteenth century.

Although the narrative of Bacon's visionary genius originated centuries after his death, the narrative positions him in relation to his own time and his future according to the logic of potentiation and supersession. Bacon's precocity grows out of the idea that the potential for the realized future exists in the past; this supposed precocity is premised on the belief that his ideas superseded those of the Arabic scholars whose work influenced him. This typological reading arises out of late antique and early medieval replacement theology, the doctrine that the Christian Church has replaced—or superseded—the Jewish people as chosen by God.[14] Replacement theology, also called *supersessionism*, shaped medieval Latin Christian theologians and natural philosophers like Bacon and remains a key doctrine of most sects of Christianity. This doctrine asserts that the present is necessarily and only a Christian present, one that has superseded Jewish temporality and excluded Jews from the present, instead keeping them relegated to "that was then."[15] The framework of replacement theology was transposed onto a chronology of scientific development by William Whewell and others. This intellectual graft, combined with nineteenth-century medievalism and historical periodization, also made possible a priority claim about modern science (that it is fundamentally European) in the service of Christian nationalism.

Bacon's historiography also provides a vantage point from which to explore the gradual erasure of Arabic texts and scholars from the history of medieval science, demonstrating how the ideology of Eurocentric intellectual superiority rests on and upholds the related ideology of Christian supremacy. Over time, Bacon's interest in and reliance on Arabic natural knowledge disappeared from academic and more popular representations of his genius. Changing views toward texts written in Arabic from the thirteenth century through the twentieth century contributed to a narrative of scientific development focused entirely on Latin Christendom as a site of intellectual innovation and scientific advancement. But, as I show, this narrative received a powerful and purposeful boost in the nineteenth century from Whewell, who suppressed Bacon's intellectual debt to his Arabic forebears to present a historical narrative that valorized European Christianity.

The Structure of Scientific Progress

In this book I identify and unpack the inventions necessary to forge the narrative of scientific progress invoked by Whewell and demonstrate how that narrative was constructed to produce and transmit an ideology of superiority that continues to this day. I explore three linked groups of necessary inventions in chronological order, from the thirteenth century through the twentieth, and analyze how they reinforce narratives of progress and Eurocentric intellectual superiority. The first of these are Bacon's inventions, necessary to his plan for spiritual enlightenment and Christian political power; namely, *scientia experimentalis* and the kinds of technologies that it relied on and made possible. Bacon's own interest in a new method for studying and understanding nature and in the inventions that *scientia experimentalis* could produce were of interest to later readers and thinkers. Bacon's writings on *scientia experimentalis* and the devices and processes that it promised formed most of the material that circulated with his name attached in the centuries after his death; these figure prominently in the next group of necessary inventions, the establishment of "the Middle Ages" as a historical period and the creation of "Roger Bacon" as a historical character. By analyzing Bacon's representation over the longue durée alongside his works, influences, and historical context, I illustrate how Middle Ages as an invented historical period is central to understanding modern science as an analytical category and the narrative of scientific development, the final necessary inventions. Thanks to Whewell and those who followed him, Bacon figures as the exceptional thinker from this period whose far-sighted vision of extraordinary machines and whose insistence on using both theory and experiential proof to confirm natural knowledge anticipated later scientific developments by centuries.[16] More importantly, characterizing Bacon in this way made it possible to claim Europe as the sole origin for modern science, a claim that has been widely accepted.

Whewell's account of the history of science, and Bacon's role in it, remained intact even as new theories about scientific progress articulated a different relationship between Christianity and science. Whewell, like Bacon, wrote from a position of belief in a divinely created universe and understood the study of natural laws and phenomena as concerning secondary causes, and believed that studying nature brings the student closer to the Christian God. By the late nineteenth century, the conflict thesis put forth by historians John William Draper (1811–82) and Andrew Dickson White (1832–1918), held that Christianity was hostile to scientific development. In their revisionist history they drew on a canard about Bacon, barely supported in the historical record, that he had clashed with religious authorities and been condemned and imprisoned for his beliefs.[17] Bacon figures as an important example in White's *History of the Warfare of Science with Theology in Christendom* (1896). "The first great thinker who . . . persevered in a truly scientific path. . . . More than three centuries before Francis Bacon advocated the experimental method, Roger Bacon practiced it. . . . [In] an age when experimenting was sure to cost a man his reputation, and was likely to cost him his life, he insisted on experimenting, and braved all its risks. Few greater men have lived."[18] In White's representation, Bacon is a freethinking rationalist risking social opprobrium, his liberty, and even his life to pursue knowledge in the face of resistance from the Church. Yet even in this version Bacon remains a messianic figure who reveals the potential latent but inherent in Western culture until modernity made widespread progress possible. Regardless of whether he is presented as an orthodox philosopher or a heterodox maverick, each "Roger Bacon" does the same work in the larger scheme of progress, thereby demonstrating the necessity of "Bacon-as-visionary."

Three themes emerge from charting these necessary inventions and the ideological edifice they support. The first theme concerns the role of the Middle Ages in the narrative of scientific progress. Roger Bacon was a product of his context, and his work is in many ways representative of his intellectual background and religious vocation. Bacon's interest in *experimenta*, including observation and proof, and in the utility of knowledge, reflected his larger interest in how the careful, sustained study of nature could bring people to Christianity, and could bring Christians closer to attaining *sapientia*, a state of learned wisdom and ecstatic communion with God.[19] Bacon's deep familiarity with Aristotelian natural philosophy led to his interest in extending lifespan using prepared dragon flesh, and his belief in the imminent arrival of the Antichrist and the need to strengthen Christendom in response drove much of the urgency behind his work on *scientia experimentalis.*[20] Besides understanding Bacon's medieval context as productive for his thinking, this context also produced the replacement theology that governs the logic of our contemporary historical periodization. That periodization includes the invention of the medieval period itself as a historical era marked in opposition to modernity and as a devolution from antiquity.

Accessing Bacon's own historical context makes it possible to apprehend the gradual erasure of Arabic texts and scholars from the history of science, highlighting the second theme of this study: how science as a knowledge practice and an analytical category functions within the mutually supporting ideologies of Christian supremacy and Western intellectual superiority. Multiple, overlapping trends in intellectual culture beginning in the early modern period began to question the accuracy of earlier Arabic translations of Greek texts, to Orientalize Aristotelian natural philosophy produced in Arabic within Europe, and, eventually, to characterize Arabic science in the medieval period as primarily concerned with translation and conservation, rather than innovation or experiment. As I show, Whewell is the inflection point for this narrative, which was enthusiastically taken up by later scholars in both French and English history and philosophy of science in a context of colonial projects in North Africa, central Asia, the Arabian Peninsula, and the Indian subcontinent.[21]

The third major theme I explore in this book is how exceptionalism and exemplarity operate in knowledge-making and the transmission of ideology. Bacon's representation as exceptional for his time and exemplary of scientific thinking indicates how these concepts uphold historical periodization and existing ideas about supremacy and value. As such, this book is necessarily about the history of scholarship, because the history of scholarship offers insights into how knowledge is revised, reassessed, and sifted into categories like "orthodox" and "heterodox."[22] By presenting Bacon's writing, his influences, and context alongside the major currents of his historiography I demonstrate the opportunities for knowledge-making that exemplarity fosters and forecloses.

Because the production of scholarship is a key theme of this book, I note that Roger Bacon's work has been a subject of continued interest over the last thirty years. Scholars of medieval alchemy have looked to Bacon's genuine works and those later alchemical texts attributed to him to articulate his ideas and explore his impact on the development of alchemy, while those working on books of secrets have clarified Bacon's investment in the genre.[23] The work of Jeremiah Hackett has been enormously important for understanding how Bacon conceived *scientia experimentalis*; his work in turn has spurred new and additional avenues of inquiry in the history of science and history of philosophy.[24] More recently, Amanda Power has done the same for understanding Bacon's commitment to the Franciscan Order and his faith.[25] More broadly, over the same period scholarship on medieval science has expanded beyond the archive and intellectual framework of Aristotelian commentary, and Bacon has been considered within his own context rather than removed from it.[26] Work on *experimenta* in the twelfth and thirteenth centuries has clarified the sustained interest in investigating natural phenomena and using testing to do so, while scholarship on the role of Latin translations of Arabic texts related to *experimenta*, secrets, and natural philosophy has defined the importance of those translations to the development of natural philosophy in Latin Christendom.[27] And although academic focus on medieval Arabic science has been sustained for a very long time, it is only in the same period that such scholarship has begun to be incorporated into the scholarly conversation in English, thereby correcting the previous narrative (promulgated by Whewell and taken up by many others) that scholars working in Arabic had merely translated Greek scientific texts into Arabic, preserving them until they could be translated into Latin and engaged intellectually.[28] Bacon's interest in the utility of knowledge to serve political ends correlates to widespread elite interest in "courtly science"—those branches of knowledge concerned with making interventions in the natural order, either to improve the human condition or to consolidate power (or both).[29] Furthermore, Bacon, like many natural philosophers and theologians, engaged rigorously with the ideas of scholars—from many religious and ethnic backgrounds—writing in Arabic, which were available to him in Latin translation.[30]

The production of scholarship includes the locations of knowledge-production, especially the university. Roger Bacon's intellectual formation emerged from the traditional liberal arts curriculum and in the context of the university. During the thirteenth century, the university was an institution for the assimilation and legitimation of knowledge that came from outside of a Christian context, making it compatible with Christian doctrine. Bacon both taught in the university and reimagined the university curriculum to develop his own theories, which later became hailed as an early example of scientific rationality. Although Bacon left the university context and looked beyond it, the university figures strongly in this book as the site of knowledge-making and the production of ideology, especially since the nineteenth century.[31]

Organization of This Book

I trace these three themes across five chapters, proceeding chronologically. This organization serves the sedimented and multiply constitutive nature of knowledge-production. The first two chapters explore Bacon's necessary inventions, beginning with *scientia experimentalis*: what it is, what it is for, how it intersects with theoretical knowledge (*argumentum*), how to acquire it, how to use it. It was the cornerstone of Bacon's larger project of major curricular and epistemic reform, which he articulated at length in response to the invitation from Pope Clement IV. *Scientia experimentalis* often made use of different types of devices and *apparata* and enabled the creation of new machines and the recovery of lost technologies. Bacon argued for the possibilities afforded by *scientia experimentalis* in the fields of alchemy, optics, and devices.

The second chapter places Bacon's work in context, by exploring his influences, which encompassed the learned communities of Oxford and Paris, the ideas of Arabic scholars, travel accounts, and historical chronicles, and his audiences, including secular and religious rulers. Bacon was concerned with the application of natural knowledge to political and moral Christian leadership, and in *Hidden Powers* and his three major works from a decade earlier he emphasized how *scientia experimentalis* was fundamentally concerned with epistemic utility. In these texts Bacon articulated how natural knowledge—*scientia*—intersected with the past, present, and potential future. By delving into Bacon's work on *scientia experimentalis*, his influences and imperatives, I convey some of the complexity of learning in medieval culture and provide a comprehensive grounding in Bacon's biography, milieu, and ideas to show how they were by later accounts amplified, reconfigured, or edited.

In Part II, I investigate the transformation of Bacon's context into "the Middle Ages" and his change from actor to character. Identifying what aspects of Bacon's work lent themselves to the interests of later periods, it becomes clear how early modern natural philosophers in the sixteenth and seventeenth centuries translated Bacon into a figure much like themselves, however they were. Chapter 3 reckons with the inventions of different posthumous versions of "Roger Bacon, English genius" that gradually replaced the historical figure of Roger Bacon from the fourteenth to the seventeenth century. This chapter also traces the necessary invention of the medieval period as a foil to the Renaissance and tracks the changing status of Arabic authors and Arabic texts.

Part III considers the fate of Roger Bacon as science changed from a type of knowledge into a method of knowing, and a related narrative of scientific progress emerged. Chapter 4 illustrates how scholars in the nineteenth century looked to the earlier, laudatory estimations of Bacon and amplified them. The most influential of these was Whewell, who, in his attempt to explain how scientific knowledge is made, saw Bacon's ideas on these topics as like Francis Bacon's epistemology of inductive reasoning and like his own account of scientific knowledge-creation. Before his epic *History of the Inductive Sciences* and *Philosophy of the Inductive Sciences*, Whewell wrote the definitive work on Gothic architecture in northern Europe, meaning that he was responsible for both codifying the history and method of modern science and also a cornerstone of medieval aesthetics.[32] He played an important role in fostering the medievalism of the Gothic revival of the nineteenth century by articulating an aesthetic investment in medieval Latin culture through his scholarship on architecture. Placing Whewell's thinking—enthusiastically taken up by his contemporaries and later writers—for how we understand medieval science, and, by extension, modern science, in context with late antique and medieval replacement theology elucidates the relationship between periodization and Christian supremacy.

The fifth chapter examines Whewell's legacy in the historiography of science in the twentieth century, revealing how the entrenched story of scientific development—and a specific definition of science—rests on medieval Christian theology and a fantasy of modernity. This is true even after new currents in historiography in the late nineteenth and early twentieth century, such as the conflict thesis, washed away the Christian belief—as Whewell had not—that suffused Bacon's work, revealing a new version of "Roger Bacon": the freethinker and visionary who was persecuted by the Church. This version of Bacon lent authority to the argument that religion is hostile to science and existed alongside the version of Bacon as the pious Christian and "first scientist," which in turn lent authority to the idea that science and religion have a syncretic relationship. Whewell's framing was assimilated enough that Kuhn, writing over a century later, could base his argument on the unquestioned paired assumptions that the intellectual heritage of ancient Greece is necessary to scientific thinking and that science only emerged within modern Europe. Implicit in his assertion is the periodization that we recognize as ancient, medieval, and modern. In Kuhn's formulation—one that he, like we, inherited—the Middle Ages are present but not named. They are the period in between antiquity and modernity, evident by omission and characterized by absence and deficiency. Kuhn focused on the early modern period for two of his case studies, and many others before and since have done likewise when considering when and how modern science emerged or was invented. How did heliocentrism replace geocentrism in European cosmology?[33] What happened in European intellectual and political culture between roughly the sixteenth century and the eighteenth?[34] How and why did experiment—and an experimental method—emerge as a critical tool for natural inquiry in the early modern period?[35] Behind all of these questions is an unspoken, basic, and commonly held assumption that these examples bear special need for explanation because they occur in such stark contrast to the period that preceded them—the medieval period.

Concluding with a consideration of Bacon's stature in the twenty-first century, it becomes clear that even as the scholarly conversations around Roger Bacon and his medieval context have changed significantly, the earlier, established narrative first articulated by Whewell remains intact. This narrative continues to be rehearsed, especially in popular culture about the medieval and early modern periods and in another division of the university—the natural and human sciences. The history presented in science textbooks privileges a teleological viewpoint and reproduces a Eurocentric narrative of scientific development, in which the Middle Ages remains characterized by absence and stagnation, with Bacon often as the only notable figure. A more detailed understanding of periodization will enable new perspectives on the history of science, on what constitutes "modernity," and our relationship to the past. Putting this understanding into conversation with work on the effects of colonization and its legacies on the history of science since the early modern period can yield new perspectives on the past and the present, and consequently, the future.[36]

Necessary Inventions reveals how we have inherited from the distant past a way of thinking about science, objectivity, and knowledge, and a way of thinking about ourselves in the present. The Middle Ages as a historical period and Roger Bacon as a historical character are earlier inventions that were necessary to creating a narrative of scientific development that privileged

Christian Europe as the origin for modern science. The narrative structure for understanding the relationship of the Middle Ages and Roger Bacon to the development of modern science is borrowed from early medieval Christian theology. Far from being secular, modern, and objective, modern science is Christian, medieval, and bound up with concepts of replacement and superiority. Understanding how these categories—medieval and modern, superstitious and scientific—were constructed and why makes it possible to get a different, possibly clearer, perspective on the past, and to appreciate how ideology shapes the historical narrative, how we see ourselves, and even the definition of science itself.

PART I

Chapter 1

Roger Bacon and *Scientia Experimentalis*

Therefore, all things must be verified via the route of experience.
—Roger Bacon, *Great Work*, ca. 1266

Sinister forces threatened Christendom. To know how to defeat them, Pope Clement IV (1265–68) wrote to Roger Bacon in the summer of 1266 and commanded him to "explain to us the remedies that you think we should adopt to address those issues that you have recently described as the occasion of such great danger, and do this quickly and secretly as possible."[1] Though the issues imperiling Christendom remained unenumerated in Clement IV's letter, by 1265 there were a number of possibilities from near and far that could have occasioned both Bacon's intensity of conviction and Clement IV's concern. King Louis IX of France (r. 1226–70) had presided over the expensive and catastrophic losses of the Frankish Christian troops in Egypt and had been taken captive during the Seventh Crusade (1248–54). This campaign had been launched in response to the psychologically devastating recapture of Jerusalem by the Khwarazmian Turks in 1244, which saw Christian sites desecrated and the Christian population slaughtered.[2] And a new nation threatened the borders of Christendom and the balance of power: Since the 1240s the Mongols had established themselves in the Balkans and Ukraine, and by 1260 had defeated the 'Abbasid Caliphate, based in Baghdad, and the 'Ayyubid dynasty, based in Aleppo and Antioch.[3] Bacon wrote about the threat the Mongols (Tartars) specifically posed to Christendom, allied to the Antichrist, itself a harbinger of the end times: "In addition, the Tartars have undoubtedly already torn down these gates, overrun the prison of Alexander, and laid waste to the world. Therefore, we must be very afraid of them."[4] By Bacon's time the legend that Gog and Magog, along with the Jews and other peoples, had been imprisoned by Alexander the Great, and which dated back to at least the first century of the Common Era, had become folded into the lore of the Antichrist. Bacon believed that the Mongols had released the nations that Alexander the Great had imprisoned, along with Gog and Magog, and that this inimical host would come from the east, working in tandem with the Antichrist (coming from the west), to destroy Christendom.[5]

The time seemed out of joint closer to home, too, where the natural order of things was upended. Paris, where Bacon had lived, lectured, and researched in the 1240s and 1250s, had been seized by growing unrest and outbreaks of violence brought on by tensions over position and curriculum between the secular faculty and the mendicant faculty (who were either Franciscans or Dominicans).[6] After the shocking—and humiliating—defeat of the forces of Louis IX and the king's capture in the Seventh Crusade, a mass movement of anticlerical French peasants, following a charismatic leader, surged through Paris in 1251 attacking the clergy.[7] The southern part of France had been wracked, for decades by this point, by an internal crusade against groups identified as heretics, in what was called the Albigensian Crusade. The French king, Louis IX, extended his concern about internal threats to his kingdom to include the large and prosperous Jewish community, culminating in the condemnation of the Talmud, new restrictions on the Jews living in his kingdom, and a public bonfire of Jewish books in 1241.[8]

England, Bacon's home, was likewise in the grip of internal political strife between the factions of the king, Henry III (r. 1216–72), and the barons who opposed him. This ongoing conflict ultimately culminated in what became known as the Second Barons' War (1264–67), and the capture of both Henry III and his son and heir, Prince Edward (later King Edward I) by anti-royalist forces. As a result, England was placed under interdict, a collective penalty meted out by the Catholic Church that excluded the faithful from virtually all the sacraments. As was the case in France, these internal political tensions also found expression in a brutal series of anti-Jewish pogroms.[9] Bacon mentioned the war in England, along with the violence in Spain and Italy, and the general air of discord as due to the malign influence of a comet under the sign of Mars.[10] The war in England had more immediate impact on Bacon, as well, in that his family, supporters of King Henry III, were impoverished and exiled after the victory of the anti-royalist barons at Lewes in 1264.[11] From Bacon's perspective it was easy to believe that things were bad and getting worse. Nor was he alone in thinking so. "And it is believed by all who are wise that we are not far removed from the times of Antichrist."[12] And the Antichrist would use every advantage, including exploiting the hidden powers of nature.[13]

To counter these threats and others, Bacon articulated a unification of natural philosophy and Christian metaphysics in what became known as the *Great Work* (*Opus maius*), the *Lesser Work* (*Opus minus*), and the *Third Work* (*Opus tertium*), which he sent to the pope between 1266 and 1268. In the *Great Work* Bacon clearly articulated his views about the central importance of "the route of experiment" (*via experientiae*) to understanding the natural world and, ultimately, nurturing a closer connection to God. *Scientia experimentalis*, which Bacon outlined in the sixth and penultimate book of the *Great Work* as well as in other works from his maturity, is both a branch of knowledge and a method for knowing, similar to logic.[14] Proper attention to natural knowledge from the senses (*experimentum*) could confirm theoretical knowledge or knowledge from first principles (*argumentum*), put received wisdom to the test, and could yield greater understanding of natural phenomena, including particular or inexplicable phenomena, like magnetism. This deep understanding of phenomena and causes, combined with human ingenuity and know-how, could create machines, effects, or processes and yield new knowledge about the natural world.

Both *scientia experimentalis* and the new processes and devices it promoted are the first necessary inventions under consideration, and they were necessary to Bacon's argument for epistemic and curricular reform. Often misleadingly translated as "experimental science" and more accurately understood as natural knowledge gained through concentrated sense experience, *scientia experimentalis* authenticated natural knowledge and offered a method beyond trial-and-error for human ingenuity to harness the power of nature. Greater familiarity with natural knowledge would lead to a greater communion with God, because, as Bacon (and many others) believed, all human knowledge was of divine origin, revealed by God to the earliest prophets.[15] Bacon believed in the fundamental unity of the sciences: that each branch of knowledge was necessary to attaining *sapientia*, or wisdom, and that each branch was part of the larger tree of theology.[16] According to Bacon, next to the study of theology itself *scientia experimentalis* provided the most important step toward *sapientia*, by which Bacon meant both ecstatic communion with God and the state of learned wisdom that could be harnessed to serve Christianity.[17] *Sapientia* became associated with the Franciscans by Alexander of Hales (ca. 1185–1245), like Bacon an Englishman from the prosperous gentry who entered the Franciscan Order in midlife.[18] A generation older than Bacon, Alexander was of the first group of Latin Christian scholars to grapple with newly available Aristotelian texts and the many responses, syntheses, challenges, and commentaries to Aristotle that accompanied them. Like Bacon (and many others) he viewed the thorough, systematic understanding of nature as a path to a closer connection with God. Bacon, influenced by Alexander of Hales and other important early Franciscans, like Robert Grosseteste (1175–1253), argued that *scientia experimentalis* could help the devout Christian come to understand the truth of Scripture and come to a closer understanding of God. "In truth, I demonstrated this property, that according to this branch of knowledge [*scientia experimentalis*], which is next to moral philosophy the most useful, and will present the literal truth of Scripture, so that through suitable adaptations the spiritual sense may be derived."[19] This close relationship between *scientia experimentalis* and *sapientia* and the importance of the former to the goal of a profound connection to God organizes Bacon's *Great Work*, in which the sections on *scientia experimentalis* and moral philosophy close out the work as the sixth and seventh parts, respectively. Thus, mastery of *scientia experimentalis* served the personal—individual character and Christian devotion—and the political—strengthening Christendom. Over the course of this chapter, I will outline why Bacon considered these inventions necessary; how *scientia experimentalis* figures into Bacon's plan for curricular reform; and the main examples of what mastery of *scientia experimentalis* might accomplish: optical devices, alchemical processes, and machines.

The Case for Reform

Part of the challenge of meeting the many existential threats facing Christianity, according to Bacon, was that the intellectual culture was inadequate to the task. For Bacon, education was a matter of morality: what you learn and how you learn can lead you toward a fulfilling and good life, or away from that life. And because natural knowledge was divine in origin, becoming educated to understand nature allowed one to live a fulfilling and good life.[20] Moreover, natural knowledge properly ordered as part of moral philosophy would yield benefits to the polity as well as the individual. "But all the works of art and nature are directed to these moral works and exist for their sake; and they have no use except insofar as they serve the work of moral precepts. . . . And this science [of moral philosophy] is the mistress of all the parts of philosophy and uses and commands them for the benefit of cities and kingdoms."[21] Not only would *scientia experimentalis* lead to new technologies, greater Christian piety, and a stronger Christendom, but it was a powerful tool to diminish the advantage of the Antichrist and his minions.

Yet the curriculum at the universities fostered the wrong habits of mind and neglected important subjects, such as languages, optics, and, of course, *scientia experimentalis*. Throughout the works that Bacon sent to the pope, he expounded on the widespread ignorance of the *vulgus*, who, even if they were educated, were still unlearned. The four fundamental errors that made true understanding impossible were, according to Bacon, "the example of weak and unworthy authority, the durability of custom, the ignorance of the masses, and the concealment of one's ignorance with an apparent display of wisdom. Everyone is implicated in all of these, every rank beholden. People of every station use the three worst arguments to come to the same conclusion, namely, this is the example of our authorities, this is the custom, this is popular; therefore, it holds true."[22] People blindly believed any "authority" of sufficient antiquity or popularity; could not admit their ignorance; and refused to examine, let alone update, their priors. Slavish devotion to authority or popular opinion was not the only impediment to true knowledge. Overreliance on using dialectic—logical reasoning, or *argumentum*—to the exclusion of other epistemic tools kept people learning in the wrong way. Bacon rejected the method of studying theology based on the *Sentences* (*Sententiae*), in which conflicting readings of Christian scripture and commentaries by the early Church fathers were reconciled through dialectic. This is probably why Bacon did not write his works as disputations or *quaestiones*, two genres that commentaries on the *Sentences* often took, but rather as treatises.[23] All this focus on learning the wrong things and in the wrong way left no room for new subjects, like optics and *scientia experimentalis*.

Bacon was intimately familiar with the curriculum in the universities. Like his contemporaries Vincent of Beauvais (ca. 1190–1264), Albert the Great (ca. 1200–1280), and Thomas Aquinas (ca. 1225–74), he belonged to the two great thirteenth-century institutions of Latin Christendom: the university and the Mendicant Orders. He studied at both Oxford and Paris, had lectured at the latter and possibly the former, and then left the university to pursue a period of independent study for almost a decade before entering the Franciscan Order. Everything we know about Bacon's life comes from the asides and snippets he included throughout his writings. He was from a wealthy family (although we do not know which one, as Bacon was a common surname in England and Normandy at this time), and born in the second decade of the thirteenth century.[24] The year of his birth has been suggested either as 1214 or 1220, based on a statement he made in 1268: "I have labored for many years in the sciences and languages, and I have now dedicated forty years to them since I first learned the alphabet."[25] Bacon was educated in keeping with the customs of the Anglo-Norman elite, beginning his formal education early, probably in private instruction.[26]

Bacon likely continued his education at Oxford in the early 1230s. Evidence for his education and location during this period is patchy and based largely on passing references in his own works to scholars such as Robert Grosseteste and Adam Marsh, whose presences in Oxford (and elsewhere) have been concretely established.[27] Certainly, during this period Oxford offered more opportunities to study Aristotelian texts in the natural sciences than Paris, which was under stricter ecclesiastical oversight.[28] And it was the study of Aristotelian natural philosophy that drew Bacon's interest. "I heard the works [of Aristotle] more diligently and read them more than the others."[29] The arts curriculum in which Bacon was formally educated and which he later taught had its roots in classical antiquity. This course of study was a conglomeration of the remnants of the Pythagorean curriculum (the *quadrivium*, comprising the subjects of arithmetic, geometry, astral science, and music) and rhetorical training (the *trivium*, comprising grammar, rhetoric, and logic).[30] Due to this accretion of subjects the curriculum had difficulty accommodating new subjects (a problem since antiquity), because it lacked a unifying and clearly articulated epistemology. While at Oxford, Bacon also had access to Greek and Arabic manuscripts and to the Jewish community; these may have kindled his later interest in learning Greek, Hebrew, and Arabic.[31]

By the 1230s the restriction on studying the books of Aristotle on logic and natural philosophy in the Faculty of Arts at Paris had eased somewhat, and Bacon moved there for further study, likely in the late 1230s. He took up a career at the university there as a secular master, lecturing on the Aristotelian *libri naturales*, as well as music, mathematics, and geometry for several years in the 1240s.[32] The university of Paris tightly controlled its teaching privileges and attendant rights, so it seems likely that Bacon must have studied at Paris long enough to fulfill the criteria for the master's degree, the prerequisite to be allowed to lecture at Paris.[33] His early works reflect his educational context and cohere around the university arts curriculum, including works on Aristotelian metaphysics, physics, and natural philosophy, as well as grammar and dialectic (logic).

Universities emerged in Latin Christendom at the end of the twelfth century as places where motivated scholars could pursue deeper engagement with textual knowledge. Initially more ad hoc than the schools based at monasteries and cathedrals, universities as institutions comprised groups of faculty, organized by subject (law, theology, medicine, arts) and groups of students (often organized loosely by vernacular language or nation).[34] These were profoundly Christian institutions, closed to non-Christians, under ecclesiastical oversight and offering clerical protections to students and faculty (a part of the conflict between "town" and "gown"). Some of the knowledge under discussion was newly available, such as some texts of the *Organon* (a group of Aristotelian texts on deductive and inductive reasoning) and natural philosophy (including other texts by and attributed to Aristotle) that were translated into Latin beginning in the later twelfth century. Other knowledge was based on this Aristotelian epistemic regime but was much more recent and came from authors writing in Arabic. Whether new or newly available, this knowledge was produced outside of a Christian intellectual framework and had to be brought in line with Latin Christian orthodoxy or else be deemed forbidden. This tension appears throughout the thirteenth century as successive restrictions on studying certain works or engaging certain premises therein that were deemed heretical (for example, Aristotle's claim that "nature abhors a vacuum").[35]

By the middle of the thirteenth century, the arts curriculum comprised texts from the old curriculum combined with texts newly available, all in Latin translation, and Bacon seems to have been interested in both. The "old logic" curriculum included Aristotle's *Prior Analytics*, and commentaries on Aristotle and Plato from the late antique authors Boethius and Porphyry. The "new logic" curriculum included Aristotle's *Posterior Analytics* and the works of late-antique Latin writers Priscian and Donatus on grammar, as well as newer works, like the twelfth-century *Book of the Six Categories* (*Liber sex principiorum*), a commentary on Aristotle's work *On Categories* (part of the *Organon*) that was erroneously ascribed to the twelfth-century logician and philosopher Gilbert de la Porrée (ca. 1090?–1154). The "new logic" also included the Aristotelian *libri naturales*, or books on nature, that so galvanized

Bacon: *Physics*, *Metaphysics*, *On Animals*, *On the Heavens*, *On Generation and Decay*, *On Meteorology*, *On the Soul*, and *On the Senses*.[36] The challenges to Christian doctrine that some of these texts posed brought the Faculty of Arts (mostly composed of the secular faculty) into conflict with the Faculty of Theology (after the 1230s increasingly staffed with faculty who were also friars), yet these subjects were also often lectured upon. Bacon lectured in Paris on geometry, mathematics, and music, as well giving at least twelve lectures on Aristotle's *Metaphysics* and the *libri naturales*.[37]

Call and Response

Bacon's clearly articulated sense of what threatened Christendom and Christianity and certainty in how to overcome these threats may have commended him to Clement IV. The two men had become acquainted several years earlier when both were in Paris. By this point, Bacon had entered the Franciscan Order about a decade prior, in the mid-1250s, and had since made the acquaintance of an ambitious and well-placed French cleric, Gui of Foulques (1190–1268).[38] In 1264, shortly before his elevation to the papacy, Gui acted as papal legate to England during the Second Barons' War. Due to the conflict, he was unable to reach England and stayed in Paris (it was Gui who had placed England under interdict). Although not a Franciscan, he was an admirer of St. Francis (ca. 1181–1226) and a supporter of the friars, and he likely sought out the Franciscan foundation while remaining in Paris and encountered Bacon there.[39] In early 1265, while still in Paris, Gui was named pope and returned to Rome, but the two men remained in contact. At some point in the months earlier, Gui had asked Bacon to send him his completed, comprehensive work of philosophy, which Gui believed had already been written. This letter and its response have been lost, although both Bacon and Gui refer to them.[40] "We have gratefully received your letter of devotion: and we have carefully attended to the words, which for the explanation of them our beloved son Sir William, called Bonecor, related to us in his own voice, both faithfully and prudently."[41] Bonecor was highly placed, as Henry III's long-serving envoy to the papal court, and had also spent time in Paris, and he was clearly educated enough and familiar enough with Bacon's ideas to be able to explain them "faithfully" to the pope.[42] Interested by what he had learned from Bacon's response, Clement IV replied with a second request in July 1266 to receive "a fair copy" of Bacon's comprehensive work, and commanded him to "do this as secretly as you are able to."[43] This second request had the authority of a papal mandate, impossible to ignore or refuse.

Though gratifying, the pope's command put Bacon in a tough spot. Pope Clement IV's request explicitly set Bacon at odds with the rules of the Franciscan Order and placed him under financial pressure. As a Franciscan, and sworn to a life of poverty, Bacon had no funds of his own to produce the work Clement had asked for, and his own family was recently impoverished. As he later reported back to the pope, "I therefore sent to my brother, a rich man in my country, who, standing on the king's side, with my mother and brothers, and the whole family, fled into exile, and having been several times captured by his enemies, redeemed himself with money; and therefore destroyed and impoverished, he was unable to help me, and even to this day I have had no answer from him."[44] Furthermore, there were restrictions in place that forbade Franciscans from sharing their work with anyone outside the Order without permission, which meant that if Bacon were to obey the pope's command for secrecy, it would put him at odds with his Order.[45] And Bacon seems to have regretted whatever he said to the pope that led him to believe that Bacon's work existed as a complete and polished text, rather than in partial drafts and notions. "As God is my witness, when I told Your Holiness that I was ready, I meant that the writings were to be made, not already made."[46] (Relatable.) Essentially, Clement IV commanded Bacon to send him a copy of a book he believed already existed, without any funds to carry out this task, while keeping the entire project from his superiors and putting him at risk of censure. The urgency and anxiety Bacon faced are palpable at moments throughout the *Lesser Work* and the *Third Work*. One historian has described Bacon's state of mind when he wrote the *Third Work* as "a little unstrung," due to the pressure he faced.[47]

Nevertheless, to fulfill the pope's command Bacon sent him, in batches, a combination of previously completed works, and several very long treatises. The three main works, the *Great Work*, the *Lesser Work*, and the *Third Work* "were preliminary writings, meant as pleas for help and financial support in reaching his real objective: the composition of a fundamental work about all the existing sciences: namely, his dreamed-of *scriptum principale*."[48] He continued to rework and refine his ideas about *scientia experimentalis* in later works (several discussed below), but never completed his *magnum opus* on the unification of knowledge under the organizing principle of Christian theology. Bacon's surviving corpus is a combination of polished, completed texts along with extensive (and sometimes recursive) drafts, which were then later copied piecemeal and often circulated as stand-alone texts. Because of the nature of his corpus it has been difficult to determine how his ideas developed and the order in which he wrote some of his works. Nevertheless, he seems to have had the text that became known as the *Great Work* to hand, as he produced it fairly quickly for the pope.[49] This was intended to be a preliminary draft to get financial support from Clement to carry out the remaining work; as he explained to the pope, he needed assistants for help with research, languages, and making diagrams to bring the work to fruition.[50] Over roughly the next two years Bacon wrote what became known as the *Lesser Work* and the *Third Work*, and sent these, in batches, along with his earlier treatises on optics, *On the Multiplication of Species* (*De multiplicatione specierum*) and *On Burning-Mirrors* (*De speculis comburentibus*), a treatise on celestial divination and judicial astrology, and four separate treatises on alchemy.[51] His zeal for reform extended to the Christian calendar and historical chronology. As an example of how these were necessary to the pursuit of *sapienta* and required a rethinking of the established curriculum he also sent to Clement IV a Hebrew calendar manuscript. Understanding how the Jews reckoned their calendar (a lunar calendar with thirteen months) was necessary to know when events recounted in the Hebrew Bible had occurred. Furthermore, this kind of knowledge could then be used by Christians to bolster their efforts to convert the Jews.[52]

He made excuses to the pope for the delay in sending his completed work and wrote of his painful financial difficulties. Trying to get the necessary funds to carry out his work while not being able to discuss what he was working on, or for whom, put Bacon in an impossible situation. "I troubled many and great people, and some whose faces are known to you, but not their minds, and I told them that there was a certain undertaking of ours to be done which had to be done in France by me, but which I was not allowed to describe, and the completion of which would require a great deal of money. . . . Even my friends did not believe me, because I could not explain the matter to them."[53] Only a few helped him, by selling their own meager possessions and even borrowing money.

The injunction to keep this enterprise secret may have been due to some of the subject matter that Bacon was dealing with, easily misunderstood or potentially heretical, such as alchemy and astral prediction.[54] The alchemical treatises were sent separately and intended only to make sense when read alongside one another. Two were incorporated into the *Lesser Work*, one was included with the *Third Work*, and one is no longer extant. The treatise on judicial astrology is no longer extant or has not yet been identified.[55] Yet it seems that Bacon did confide his charge to others, at least somewhat. He writes highly of the students and colleagues that read his work, the scribe who wrote part of the *Third Work*, and that he had confided the manuscripts to the care of a student, John, who was familiar with the works and who carried them to the papal curia.[56] Furthermore, given the impossibility that Bacon could have done all this without the knowledge of his superiors in his Order, it seems that he had their support; certainly, they did not prevent him from conveying his works to the curia.[57] Although there is no evidence that Clement IV read the works he had requested before his death in November 1268, Bacon's treatises sparked interest within the papal curia, possibly because of Clement IV's interest in his ideas.[58]

Scientia Experimentalis and the New Curriculum

Fears of imminent apocalypse drove Bacon's sense of the pressing need to adopt a new curriculum and a new method of knowledge acquisition. He urged a complete overhaul of the university curriculum. And because *how* we know is as important as *what* we know, Bacon argued for the necessity of understanding the science of visual perception (*perspectiva*) and the method of *scientia experimentalis* and elevated them on par with more traditional subjects of study, like logic, mathematics, and moral theology. He articulated a natural philosophy that relied on experience working in tandem with theoretical, text-based knowledge, and that would yield useful knowledge. Moreover, studying texts and learning through syllogism and dialectic (*argumentum*) alone would not confer authoritative knowledge; experiential knowledge was required for certainty.

Because *scientia experimentalis* was so central to the political and moral strength of Latin Christendom, as well as to Bacon's philosophy, he discussed it often throughout his works. It is a method for the practical and theoretical sciences, similar to logic, enabling the practitioner to discern magic (fraud) from art and science (truth). "This branch of knowledge alone, therefore, knows how to test perfectly what can be done by nature, what by the effort of art, what by trickery; what the incantations, conjurations, invocations, deprecations, and sacrifices that belong to magic desire and dream of; and what is in them, so that all falsity may be removed and the truth alone of art may be retained. This science alone teaches all the insanities of magicians, that they may not be confirmed but rejected, just as logic considers sophistical reasoning."[59] Yet, continued Bacon, because the universities did not teach this method of knowledge acquisition "*scientia experimentalis* is completely unknown to the common mass of students" and to their masters.[60] This ignorance of method could lead to errors of reasoning that might be heretical, thereby endangering body and soul; this ignorance could also render the unwary an easy mark for unscrupulous fraudsters and magicians. Although *scientia experimentalis* might produce effects that could resemble—or surpass—those of the magicians, the effects had different causes. *Scientia experimentalis* was doctrinally sound and an important method to attaining *sapientia*, grounded in a deep understanding of natural laws and phenomena; according to Bacon, magic was caused by demons and used by people bent on deception and control.[61]

Producing wonderful effects or devices or gaining new knowledge through *scientia experimentalis* were not its only purposes. Bacon divided *scientia experimentalis* into three purposes, or dignities: 1) to confirm or refute theories and conventional wisdom; 2) to create instruments or machines to pursue knowledge; and 3) to uncover the secrets of nature and unite past, present, and future knowledge into a single, comprehensible strand. Bacon argued that sense experience was critical to understand the natural world and use that knowledge to intervene in the natural order. Although Bacon affirmed the importance of empirical knowledge (knowledge gained through accumulated, regular experience), he focused especially on knowledge gained through observation and participation to confirm or correct knowledge from first principles, textual knowledge, or from received wisdom. He articulated a theory of knowledge that relied on experience to ratify theoretical or text-based knowledge, and on instruments or devices in the service of *experimentum*. And in a slightly later work, also written to convince a ruler of the necessity of *scientia experimentalis*, Bacon articulated possibilities for creating machines and devices for knowledge-acquisition, warfare, espionage, and for detecting charlatans and their cheap tricks.[62] *Scientia experimentalis* made these outcomes possible in part by producing knowledge gained through the sensorium and from a wide range of perspectives to test theories, textual attestations, and lore.

In the medieval period *experimentum* and *experientia* were often used to connote a sense of active participation in knowing the world, one which encompassed proof and testing, but which also included empirical knowledge, gained through regular experience. Although these words were usually used interchangeably, *experientia* was sometimes used to suggest a more limited way of knowing (experience of singular events), while *experimentum* could sometimes be used to suggest the grasp of the principle behind particular or singular events.[63] Centuries earlier than Roger Bacon and his contemporaries we encounter evidence of specific, directed tests of natural substances. For example, Isidore of Seville (ca. 560–636) mentioned in his encyclopedia, or *summa*, several methods to test for the purity of balsam, an expensive, rare, and frequently adulterated natural substance.[64] An example from closer to home comes from baking with wild yeast (rather than commercial yeast). A baker drops a spoonful of the leaven (also called the sponge) into a container of cold water; if the leaven floats, it is ready to be used in the dough. In medieval terms, this test is an experiment. Once the remaining flour, water, salt, and other ingredients are mixed with the leaven it is left to "proof," and the increased size of the dough at the end of that period indicates that the yeast has "proved"—that it is active enough to make the dough rise.

The first purpose of *scientia experimentalis* concerns different ways of knowing things: through syllogistic reasoning—also called rationality (*argumentum*)—and experience (*experimentum*). (This is aside from divine illumination, which according to Bacon is the best way to attain certainty.)[65] But relying on syllogistic reasoning might leave room for doubt; *argumentum* does not always provide certainty, nor can it account for particular or irregular phenomena.

> There are two methods of acquiring knowledge, namely, by rationality and by sensory experience. Rationality draws a conclusion and makes us concede the conclusion, but does not make the conclusion certain, nor does it remove doubt so that the mind may rest in the understanding of the truth, unless the mind discovers it by way of experience: for there are many who have arguments about what can be known, but lacking experience they neglect these matters, and neither avoid that which is harmful nor follow that which is good. For if a man who has never seen fire should prove by sufficient reasoning that fire burns and harms things and destroys them, his mind would not be satisfied, nor would he avoid fire, until he placed his hand or some combustible substance in the fire, so that through experience he might prove what logical reasoning taught. But when he has had actual experience of combustion his mind is

> made certain and rests in the dazzling brightness of truth. Therefore, reasoning does not suffice, but experience does.[66]

Someone who has learned about the destructive properties of fire only through authoritative texts and *argumentum* is less certain and, in a fundamental way, knows less than the person who has both studied texts *and* burned themself. Socrates's students may have understood intellectually that Socrates was a man and that all men are mortal, and that, therefore, Socrates was mortal, but they gained a new understanding of that knowledge once Socrates died. *Scientia experimentalis* verifies the knowledge found in authoritative texts and conveyed through syllogism, but it does not replace that knowledge. Yet without both logical reasoning and experiential knowledge certainty is impossible. Moreover, Bacon makes clear the moral dimension to *scientia experimentalis*: those who lack this kind of knowledge cannot "avoid that which is harmful nor follow that which is good."

Bacon also clarified Aristotle on the importance of experience, arguing that experience in conjunction with syllogism confers greater wisdom than syllogism alone. "Thus, when Aristotle said that proof is reasoning that makes us know, this is understood as proof accompanied by experience, and not of the bare proof itself. As he said in the first book of the Metaphysics, it is said that those who know through experience both the reason and the cause are wiser than those who know something through experience and know only the bare truth without the cause [i.e., empiricism]. But here I speak of the expert who knows through experience both rule and cause. And such men are perfect in their wisdom."[67] Bacon relates this idea to the Aristotelian distinction between knowledge of a fact (*quia*) and knowledge of the cause of a fact (*propter quid*). According to Bacon the knowledge of the cause of the fact requires proof accompanied by experience.[68] For example, the cause of the rainbow—found in multiple guises and settings in nature (after a rainstorm, through a prism, in the spray of water from a boat or a water mill, in the feathers of a bird or the wings of an insect)—cannot be understood through syllogism alone, but must be investigated via experience.[69] To go back to our homely sourdough example, an empirical understanding of the use and cultivation of wild yeast in baking confers the knowledge that once the leaven can float in water it is ready to use in the dough. A more complete understanding goes a step further to encompass the cause of the phenomena, the reason that the leaven floats and the bread rises: the yeast produces a product that makes the leaven lighter than water. A scientific explanation of this phenomenon, in our contemporary terms, is that the yeast feeds on the flour and water in the leaven and produces carbon dioxide as a waste product; it is the carbon dioxide that makes the dough rise.

The second purpose of *scientia experimentalis* covers the useful and incredible results from the combination of human art (what we call technology) and natural knowledge. This includes processes to refine metals, such as alchemy, as well as medical recipes for longevity, and the creation of new devices. These examples all combine deep knowledge of *argumentum* (for example, geometry and physics) with experiential knowledge—hands-on making.[70] For example, understanding the science of the stars—the positions of planets, the appearance of unusual phenomena, like comets—requires the use of special instruments. This knowledge is crucial to being able to form accurate judgements or predictions about all manner of things (weather, the harvest, medicine, horoscopes).[71] Likewise, according to Bacon even the ancient authorities, such as Ptolemy, taught that the properties of celestial phenomena "are certified by the methods of viewing them with instruments."[72]

An example of a device that *scientia experimentalis* could make possible would be an armillary sphere combined with a magnet. Without knowing through experience that a magnet attracts iron, one could not envision how to put it to use or understand that the attractive and repulsive forces of the magnet might be related to the ebbing and flowing of tides, or other examples of action at a distance, and thus might be combined with an astrolabe to reproduce celestial motion accurately. Knowledge of mathematics alone would not make it possible to design or build such an instrument; it would require, according to Bacon, an "accomplished experimenter" with first-hand, embodied knowledge of the material as well as mathematical expertise. Devices, such as this one, could in turn yield new information about nature. "This instrument would be worth a king's ransom, and would stop all other astronomical instruments and clocks, and would be a most beautiful display of *sapientia*."[73] The astronomical instrument Bacon referred to would move in perfect harmony with the celestial spheres, and would also perfectly model their divinely created movement, thereby enabling a more perfect understanding with the divine creator—the kind of *sapientia* made possible by immersive, sustained, thoughtful engagement with learning and nature. However, such an instrument would also require the maker to have already attained some measure of *sapientia*, to have the knowledge necessary to make the instrument. This example—though speculative—encapsulates Bacon's argument for *scientia experimentalis*: mastery of this way of knowing, in conjunction with foreign languages, mathematics, optics, and astral science, enables the practitioner to attain the kind of wisdom that could, in turn, enable the creation of new and useful technology, which could then yield new knowledge.

In addition to making possible more useful instruments, *scientia experimentalis* fostered the realization of alchemical potential. Bacon sent several alchemical treatises to the pope and wrote about the art throughout his three major treatises, and elsewhere in his oeuvre as a process that could both refine metals and create medicinal elixirs for rejuvenation and health. His writing on alchemy became progressively more detailed and lengthier in the *Lesser Work* and the *Third Work*, which may reflect changes in his thinking as he brought alchemy into closer alignment within *scientia experimentalis*, or a deliberate tactic of dispersal in order to prevent the material from falling into the wrong hands.[74] Bacon's alchemy was unique to himself, a syncretic mix of alchemical ideas and texts from the Arabic tradition, Galenic medical theory, aligned with his own views about experience and experiment. Indeed, the empirical nature of alchemy, and its reliance on sensory knowledge (observing changes in mixture with sight, smell, or taste), may have been part of its allure for Bacon. Alchemy was not incompatible with religion; in fact, it drew from religious ritual and, in dealing with matter, dealt with matters of creation.[75] The longest section about alchemy appears in the *Third Work*, in which Bacon made clear the distinction between "speculative" and "operative" branches, as he distinguished them. Operative alchemy was more empirical, and less theorized. Speculative alchemy was the province of those few who had completely mastered the principles of alchemy and could apply them to both insensible materials like dyes and metals and to animated material, like the human body.[76] Speculative alchemists had the necessary mastery of theory and practice to make the alchemical elixir, which could balance an individual's humors and bring their body into perfect balance and, thus, perfect health. "But the faithful experimenter has considered the eagle, and the deer, and the serpent, and the phoenix, and many things to prolong life and renew youth, and he knows that these things are given to animals for the instruction of men; for which reason he devised noble ways for this, and used practical alchemy to prepare for him a body of equal complexion, so that this experimenter might use it."[77]

Speculative alchemy could also bestow not only "foresight and intellectual acumen, but also the knowledge Bacon considered the most important: knowledge of God. Bearing in mind Bacon's context, knowledge of God means knowledge of the Christian God and the truths of Christianity. Therefore, the use of the elixir produces not just a perfect body, but a perfect Christian, as well."[78] Bacon's ideas about alchemy were not taken up widely by his contemporaries or immediate successors. However, Bacon's alchemical ideas, especially the assertion that special lenses (burning mirrors) could concentrate celestial rays and redirect them onto already perfected human bodies, which could in turn transmit the benefits of those rays to imperfect bodies, finds later echoes in John of Rupescissa's (ca. 1310–62) theory of the quintessence.[79]

The third purpose of *scientia experimentalis* offers the promise of understanding the secrets of nature, including knowledge of the future. Utility was critically important to Bacon: Natural knowledge for its own sake was less important than how that knowledge could be used to defend Christian kingdoms or be mobilized to attain *sapientia*.[80] *Scientia experimentalis* could provide knowledge of how to use natural substances with hidden capabilities that could be turned to the ruler's advantage, because it realized and facilitated knowledge—of particulars, like non-manifest qualities, or of natural forces and phenomena that allowed the construction of devices and instruments—that could be authenticated and used. For example, "when properly prepared, yellow petroleum, which comes forth from rock, burns whatever it meets . . . and water will not extinguish it."[81] Explosive compounds of different minerals can produce deafening alarms, so that "neither army nor city can withstand them."[82] Here, and elsewhere, as we shall see later, Bacon argued for the possibility of inventions (and their applications) in the future because of their existence in the past. This line of reasoning subtly echoes the third prerogative of *scientia experimentalis*: to enable concurrent knowledge of the past and future, in the present. The third dignity of this science, according to Bacon, "consists in two things; namely, in the knowledge of the future, the past, and the present, and in wonderful works by which it excels in the power of making predictions the ordinary judicial astronomy dealing with predictions. . . . [This] branch of knowledge has discovered the definitions and paths by which it can easily answer every question, and . . . it can show us the forms of the celestial forces, and the impressions of the heavenly bodies on this world without the difficulty of ordinary astronomy."[83] Bacon's comments on this purpose of *scientia experimentalis* are somewhat obscure, but it is apparently similar in outcome to judicial astronomy, but without the need for accurate astronomical tables and expensive instruments that astral prediction usually required. His views on astral science, *astrologia*, and Aristotelian natural philosophy have been the subject of sustained debate, but it seems clear that he viewed the possibilities for prediction that astral science offered to be under the umbrella of natural philosophy, rather than demonic magic.

Mastering *scientia experimentalis* fostered outcomes that could be turned to political utility and to the benefit of Christendom. Bacon tantalized and assured his intended reader—Pope Clement IV and his curia—that ancient philosophers and just rulers, like Aristotle and Alexander, have already "invented impressive devices against enemies of the state, so that without iron or another weapon having to touch anything they could cut down any who would resist, and there are many such inventions."[84] Moreover, the subjects in the new curriculum, like optics and mathematics, have practical applications. Alexander the Great, because he understood the science of optics, used a mirror to turn the lethal gaze of a basilisk from his own army, where it was trained, back onto itself, "so that it was killed by its own venom."[85] The potential of *scientia experimentalis* could be harnessed in the service of fortifying Christendom and battling against the forces of the Antichrist, which Bacon and many others believed were imminent. Bacon urged the pope, "in order to spare Christian blood, the Church ought to consider the use of these inventions against unbelievers and rebels, and it should do so *especially* because of the coming perils in the times of the Antichrist, which (with the grace of God), if prelates and princes fostered inquiry and investigated the secrets of art and nature, it would be easy to face."[86] Demonstrations of the kinds of technological marvels made possible by *scientia experimentalis* could be used to convert non-Christians by making them believe in what they might not understand. "Unbelievers are obsessed by these insane things [marvels] and put trust in them and believe in them, and they believe that Christians used such means in working miracles."[87] Bacon here may have been referring to the kinds of mechanical marvels—elaborate fountains and artificial birds—reported at the court of the Mongol *qaghan* (khan) Mangu in Qaraqorum by Franciscan missionary William of Rubruck (1220–93).[88] The two men met a few years after William returned from his journey (1253–55), and Bacon relied on William's account of his travels for his grasp of the possibilities for the practical applications of *scientia experimentalis* in the service of the Church and against non-Christians.[89] In addition to providing useful devices, *scientia experimentalis*, as a method for knowledge-acquisition, made possible a more complete understanding of the Bible than that offered by the established method of reasoning via exegesis, and was "the most useful apart from that of morals."[90]

For Bacon, *scientia experimentalis* was a path to understanding the natural world more fully, and in a way that ultimately brought one into closer and deeper connection to God. Educated in the liberal arts and Aristotelian philosophy within the university context, Bacon ultimately became frustrated and disillusioned with what he perceived to be the limitations of the curriculum, with its focus on *argumentum* and syllogism, to meet the pressing challenges of his moment. Yet, although steeped in a traditional arts education and a product of the new universities of Oxford and Paris, Bacon ultimately looked beyond the traditional curriculum to pursue a deeper and more complete understanding of natural knowledge. He went beyond the confines of the classroom to explore knowledge from the sensorium and to hone his idea of *scientia experimentalis* and its potential. Some of this knowledge was extra-textual or came from people who did not write down what they knew; other knowledge came from texts that circulated outside of the university curriculum in genres that emphasized useful knowledge. Bacon was concerned with the utility of natural knowledge to political and moral Christian leadership.

Chapter 2

Useful Knowledge and Courtly Science

> I shall now at once relate the marvelous works of art and of nature—in which there is nothing magical—so that I may assign to them cause and manner, that it may be seen that all magical power is unworthy and inferior to these works.
> —Roger Bacon, *Letter on the Hidden Powers of Art and Nature, and the Invalidity of Magic*, ca. 1270s

Roger Bacon's education within the traditional curriculum acquainted him with its opportunities and shortcomings. Many of the influences and ideas that contributed to his theory of *scientia experimentalis* largely came from outside that traditional educational program. These include texts from his contemporaries in Latin Christendom, who shared his educational background and many of his intellectual interests, men like Robert Grosseteste and Michael Scot; texts written in Arabic and newly available in Latin translation from philosophers like Ibn al-Haytham (Alhazen) and al-Kindī (Alkindus); material on secrets, like the *Secret of Secrets*, ascribed to Aristotle; and historical texts, especially about Alexander the Great.

Experimenta/Experientia

Bacon may have been the only of his contemporaries to write thousands of pages discussing *scientia experimentalis*, but he was not the first to use *experimentum/experientia* or argue for its importance. Indeed, the concept of using experience to attain certainty about natural knowledge has many ancient antecedents as well as medieval Latin and Arabic adherents. Bacon cited Aristotle, Pliny, and Ptolemy as proponents and practitioners of gaining knowledge through experience.[1] He credited Ptolemy with the notion of *scientia experimentalis* in both the *Great Work* and his edition of the pseudo-Aristotelian *Secret of Secrets*, connecting Ptolemy and *scientia experimentalis* specifically to prediction and divination through the interpretation of natural signs.[2]

Furthermore, Bacon's interest in *experimentum* was of a piece with learned Latin culture in the thirteenth century.[3] Robert Grosseteste (ca. 1168–1253), natural philosopher, Oxford master, Franciscan, and eventual Bishop of Lincoln, used the concept of *experimentum* in his early work on comets (*De cometis*, ca. 1225). In it he argued that both experience and rationality are necessary to understanding comets and, by extension, other singular phenomena—those instances in which accumulated experience (*empireia*) cannot be applied. Grosseteste also used the concept of *experimentum* productively in his work on optical theory and his commentary on Aristotle's *Posterior Analytics*: external confirmation through the sensorium is required when dealing with universals that are complex and not easily separated from particulars, such as the phenomenon of the rainbow (a subject to which he dedicated an entire treatise).[4] After 1230 Grosseteste lectured regularly to the Franciscan foundation in Oxford, and his treatises on logic and optics circulated in and around Oxford and Paris. Bacon was likely not Grosseteste's student in any formal sense, yet Bacon clearly found Grosseteste's work generative for his own thinking. *On Comets* circulated in Paris in the decades after its completion while Bacon was living there, and Bacon also drew heavily on Grosseteste's commentary on *Posterior Analytics* when he lectured on the subject as part of the arts curriculum at the university in Paris in the 1240s.[5] Grosseteste began studying Greek after 1230 and worked with other scholars on translations of Greek philosophy, including Aristotle's *Nichomachean Ethics*.[6] Bacon spoke highly of Grosseteste in several of his writings, specifically praising Grosseteste's use of experience. "Master Robert . . . ignored all the works of Aristotle and their paths, but through his own experience, and in traffic with other writers and other branches of knowledge, found out and wrote more than a hundred thousand times as many things as are recounted in Aristotle's books or as can be taken from the corrupted translations of the same."[7] Blindly following textual authority at the expense of experiential knowledge prevented one from knowing very much at all.

This disavowal of Aristotle may seem surprising, yet it was not unusual in this context. Bacon addressed this need for openness to revision, especially in the conclusions of authorities. "For later thinkers always added to the works of earlier ones, and corrected many things and changed many more, as is especially clear in the case of Aristotle, who critiqued all the propositions of earlier philosophers."[8] Even the Church fathers—the authorities of early Christianity—changed their minds and disagreed vigorously with one another.[9] Natural philosophers in the thirteenth-century Latin tradition grappled with Aristotle's works on natural philosophy and logic and offered refinements or emendations to his ideas. Michael Scot (ca. 1175–1235), part of the first wave of Latin philosophers to engage the philosophical and scientific texts newly available in Latin, also engaged with Aristotle's assertions. Like Bacon, Scot was a polymath and natural philosopher, with scholastic ties to Oxford and Paris. He was also a courtier to Holy Roman Emperor Frederick II (1194–1250). Scot undertook at Frederick's urging new translations of Aristotle's *libri naturales* and Ibn Rushd's (Averroës) commentaries from the Arabic.[10] Frederick II himself was also part of this larger community of thinkers and writers who were interested in using *experientia* to correct or confirm knowledge found in authoritative texts. The emperor's interest in Aristotle's treatises on nature, especially the *Book on Animals* (*Liber animalium*), stemmed in part from his interest in his extensive menagerie and collections of hunting animals, including cheetahs, dogs, and raptors—hawks and falcons.[11] In the early 1240s he wrote a treatise on falconry, *The Art of Hunting with Birds* (*De arte venandi cum avibus*), and explicitly cited Scot's translation of Aristotle and accompanying commentary, and the importance of experience to correct Aristotle's mistakes.

> We discovered by experience that the deductions of Aristotle, whom we followed when they appealed to our reason, were not entirely to be relied upon, more particularly in his descriptions of the characters of certain birds. Therefore, we do not follow the Prince of Philosophers in everything; He was ignorant of the practice of falconry—an art which to us has ever been a pleasing occupation, and with the details of which we are well acquainted. In the *Book on Animals* we find many quotations from other authors whose statements he did not verify [by experience] and who, in their turn, were not speaking of what they had seen. Entire conviction of the truth never follows mere hearsay.[12]

Although Bacon does not mention Frederick's work, he was, through their shared connections in Paris, likely familiar with Michael Scot's translations of Ibn Rushd and Aristotle. Furthermore, Bacon echoed Frederick's sentiments for the need to correct Aristotle with experience, and the need for experience to test hearsay or textual authority. Frederick asserted the value of hands-on practice and experience in hunting with birds; the second half of the treatise is dedicated to practical instruction, with detailed directions for how to train and care for different raptors. Highly educated and a powerful ruler, Frederick valued natural knowledge for how it could be used—in this case, for falconry: "Utility is of the greatest importance."[13] Far more than an idle pastime, falconry served aesthetic, social, and political purposes for the elite rulers of Latin Christendom, and communicated status, wealth, and authority to other members of the lay elite and to those under their rule.[14] Bacon would echo this extension of natural knowledge for political utility twenty-five years later for Clement IV.

Other scholars with ties to Paris shared Bacon's interest in *experientia* as well as his willingness to look outside the university for knowledge. William of Auvergne (ca. 1180–1249) was Master of Theology at the University of Paris when Michael Scot arrived there to lecture on Aristotle, and he later became bishop of Paris (and therefore head of the university) while Bacon was there. William wrote extensively on natural philosophy and natural particulars and used the term *experimentores* to describe people extremely learned in natural phenomena. These individuals could discern hidden (occult) phenomena—known only through experience—such as the inherent powers of certain gemstones. "In parts of India . . . natural magic can particularly flourish there. And it thrives on account of the many experimenters there, and people who make marvelous things through skill of this sort."[15] At this point it was believed that many natural objects with hidden essences came from India, therefore it followed also that the men who were the most skilled in this kind of knowledge were found there. William of Auvergne's interest in empiricism and his role as head of the university at Paris indicate the extent to which experimental thinking was an important part of Scholastic culture in the thirteenth century.

Throughout his works Bacon confirmed the impression of a robust and sustained interest in a community of scholars and practitioners in *experimentum* and *experientia* inside and outside of the university context. In a work on the role of the sensorium in knowledge, *Common Principles of Nature* (*Communia naturalium*, 1260s) Bacon suggested that many natural philosophers, frustrated or confounded by poor translations of Aristotle and commentaries, had turned to *experimentum* to provide clarification and open new avenues of inquiry. "There are some men . . . seeing that they could not know natural philosophy through the work of Aristotle and [Averroës] his commentator [because of faulty translations], are turning to the seven other natural sciences, to mathematics, and to other authors of natural philosophy . . . and so are coming to an acquaintance with natural things, about which Aristotle, in his common books, and his expositor [Averroës], are unable to satisfy."[16] One of these natural philosophers, Albert the Great (ca. 1200–1280), like Grosseteste and William of Auvergne a member of both the university and the Church, echoed the importance of *experimentum* when investigating minerals, gemstones, animals, and plants.[17] Far from being a singular figure, Bacon's interest in *experimentum* and empirical knowledge was reflective of his immediate intellectual formation and the broader established intellectual culture.[18]

Although many of the men interested in *experimenta* whom Bacon read and perhaps met were also part of university culture, he ultimately looked outside of the university setting to pursue his interest in *experimentum*. Between roughly 1247 and 1257 he pursued a self-funded career as an independent scholar. He spent some of this time in Oxford and some in Paris, as part of a group interested in experiment, and during this time he engaged Latin translations of Arabic texts on instruments and experience.[19] He also spent some of this time studying languages, especially Greek, Hebrew, and some Arabic.[20] He insisted on the importance of knowing these languages, as well as Chaldean, for a number of reasons: "The thrust of the matter is this: a man needs to know how to read Greek, Hebrew, etc."[21] One reason is that the wisdom of the ancients was conveyed in these languages or translated into them, another was because Latin might not have a vocabulary that could accurately convey new ideas or new phenomena. "It must be considered that the translators did not have terms in Latin for translating the sciences, because they were not first composed in the Latin language."[22] An anecdote from Bacon's time lecturing in Paris vividly illustrates why Bacon thought languages were so important. While lecturing on the pseudo-Aristotelian *On Plants* (*De plantis* was believed to be a genuine work of Aristotle at this time), Bacon related a fact about the plant henbane. The Latin translation of *On Plants* available to Bacon had used the word "belenum" for henbane, and Bacon thought that this was the Arabic term for henbane, substituted by the translator. He was corrected publicly—and apparently with some mirth—by his Spanish-speaking students, who informed him that "belenum" is not an Arabic term but rather a colloquial Spanish term for the plant. Furthermore, with their help Bacon learned that the translators working in Spain had used a number of Spanish terms in their translations from Arabic into Latin, which had been taken by Bacon and many others to be Arabic terms.[23] Learning the languages in which works were first written made it possible to reduce one's reliance on potentially faulty or incomplete translations, thereby falling into error, or even heresy. For Bacon, who often complained about the ignorance and inaccuracies of translators, wisdom required careful, knowledgeable, dedicated experts, including experimenters and translators.[24]

In this period of the 1240s and 1250s, before entering the Franciscan Order, Bacon also looked to Arabic works that were now available in Latin translation. Several scholars have noted that Bacon seems to have been strongly influenced by the interest in Arabic texts that characterized Oxford in the early thirteenth century, and by the example of his fellow Englishman Adelard of Bath (ca. 1090–1155), who had traveled to Sicily, Greece, and southern Italy, and had there read Greek works in Arabic translation and newer texts in Arabic.[25] Bacon drew on a work by al-Kindī (ca. 800–870 C.E. / 185–256 A.H., Alkindus in Latin), *On Perspectives* (*De aspectibus*), which had been translated into Latin in the late twelfth and early thirteenth centuries.[26] Al-Kindī's work was a refinement of Euclid's *Optics*, based on a geometrical explanation of visual perception. Following Euclid (and Ptolemy) and disagreeing with Aristotle's theory of vision, al-Kindī posited a theory of emission, or extromission: visual rays emitted from the eye hit objects in the cone of vision, which allows us to perceive them with our eyes. Furthermore, al-Kindī argued that visual rays and rays of light act the same, and used mirrors, lenses, and other devices, like special lamps, to prove this.[27] Al-Kindī's realization that astral science and optics investigate the same phenomena—the radiation of light—informed Bacon's theory of the emanation of force, or *species*, as did his use of measurement, observation, and instruments to confirm theoretical precepts.[28]

Al-Kindī's work greatly influenced another Arabic scholar whom Bacon also considered an authority: Ibn al-Haytham, known as Alhazen or Alhacen in Latin. In his work on optics, *Book of Perspectives* (*Kitāb al-Manāẓir*, ca. 1030 C.E. / 421 A.H.), Ibn al-Haytham posited a theory of visual perception that could stand on mathematical, physical, and physiological grounds, allowing for a unification of knowledge from multiple disciplines.[29] Ibn al-Haytham disagreed with al-Kindī's (and others') adherence to emission theory, but he took from al-Kindī the idea that light rays and visual rays operate according to the same principles, therefore *sight* and *light* are part of the same branch of knowledge.[30] This perspectivist approach strongly influenced the Latin optical tradition. Ibn al-Haytham also championed the importance of inductive reasoning and testing; Beginning with induction, rather than first principles, in the *Book of Perspectives* he cited no authorities and based the argument on mathematical, experimental, or inductive grounds.[31] He wrote, "Moreover, everything we have discussed can be tested by experiment so we will attain certainty about it."[32] Using this experimental method, Ibn al-Haytham advanced a theory of intromission (that we perceive objects because they emit rays to the eye). He relied on Ptolemy's model of the visual cone, but rejected the extromission theory, reversing the direction of the visual cone and replacing visual rays with light rays. Because his theory had to reconcile existing knowledge of the behavior of rays and angles (geometry) with the physiology of the eye, and the perception of visual phenomena, he insisted on the importance of experience and demonstration, referencing experimental confirmation throughout his work and specific instruments.[33] Yet even this empirical focus and interest in demonstration was not new to Ibn al-Haytham but strongly based on Ptolemy.[34]

The *Book of Perspectives* was a masterful synthesis of natural philosophy and visual ray theory, and it seems to have been more influential in the Latin tradition than in the Arabic. It exists in one complete copy in Arabic, a handful of partial copies, and as added to part of a later work by Kamāl al-Dīn al-Fārsī, *Clarification of Optics* (*Tanqīh al-Manāẓir*).[35] The *Book of Perspectives* was translated into Latin at the start of the thirteenth century, likely in Spain, and titled *On Visual Perception* (*De aspectibus*). The work is technical and advanced, and was translated by competent translators, and likely found an interested group of readers given the expense and effort involved in translating a text of that complexity. *On Visual Perception* survives in eighteen manuscripts, fifteen of which are from the first century after its appearance in Latin.[36] Yet by 1250 optics was still not part of the arts curriculum and was only taught on occasion. According to Bacon, *On Visual Perception* had been taught only at Oxford before 1270, and only twice, so it seems likely that Bacon himself was first introduced to this work during his period of independent study.[37]

One of the earliest natural philosophers to engage with Ibn al-Haytham's work, Bacon grappled with the theory of intromission first in his work on visual perception and the emanation of force, *On the Multiplication of Species* (*De multiplicatione specierum*, ca. 1266) and in later works, especially the fifth book of the *Great Work*. Put simply, Bacon's theory of *species* states that an invisible force emanates from all objects; this emanation of force is what allows for visual perception of objects. However, Bacon also allowed for extromission, arguing that sensory perception is not passive.[38] Bacon cited Ibn al-Haytham (as Alhacen or Alhazen) as a major influence throughout the *Great Work*, in some cases adopting Ibn al-Haytham's ideas, terminology, and theory unchanged as his own.[39] Bacon synthetized Ibn Sīna's (Avicenna ca. 980–1037 C.E. / ca. 369–428 A.H.) cognitive theory of perception with Ibn al-Haytham's operational explanation. Visual perception was the most important of the senses to *scientia experimentalis*—observation usually meant visual observation—and, according to Bacon, visual perception involved reasoning. Therefore, the science of optics was the most important of the sciences, because it established the process for all other experiential knowledge.[40] Furthermore, Bacon held, following Grosseteste, that vision could lead to *sapientia*, and that the study of optics could be not just a metaphor, but a model, for attaining spiritual illumination.[41]

In addition to influencing his theory of visual perception and *species*, Ibn al-Haytham's work helped inspire Bacon's concept of *scientia experimentalis*, especially the second function, which proposed the creation of instruments to investigate nature.[42] Ibn al-Haytham introduced a new concept that diverged from Aristotle's notion of *empireia* (accumulated experience). *Iʿ tibār*—rendered in Latin as *experimentatio*—encompasses a range of meanings, including "careful consideration," "to observe carefully," "to gain experience/learn from," and "to acknowledge or recognize a quality in something."[43] The term is from the seventh verb conjugation pattern, or *wazn*, a form that implies reflexive action; in this case, observing something and taking a lesson from it. Ibn al-Haytham used this term to refer to a test to investigate physical properties directly, "with the aid of an experimental apparatus especially designed for the purpose. And the aim of the proof was to bring certainty or exactness and precision to an observation by subjecting it to an artificial situation in which the conditions could be varied."[44] This concept of proof is distinct from *empireia* and requires careful observation and measurement. For Ibn al-Haytham, *iʿtibār* was a process for confirming or disproving knowledge, rather than discovering new knowledge. Bacon echoed this concept both in the idea of using *scientia experimentalis* to confirm theories and the importance of specific instruments to do so, and he also extended this concept, suggesting that *experimentatio* could be employed to discover new knowledge as well.

Bacon also followed the works of the Arabic authorities in terms of understanding planetary influence; that is, in the science of the stars, which included observation and prediction. Celestial bodies radiated influence to earth and this influence could be measured and calculated with specific instruments and geometry. In attributing the Second Baron's War and the dire situation in England to the influence of a planet in the sign of Mars, Bacon echoed the work of Persian astronomer Abū Maʿšar's (787–886 C.E. / 171–272 A.H., known as Albumasar in the Latin West) *Book of Religions and Dynasties* (*Kitāb al-milal wa-ʿ-duwal*). This work of historical astrology provided a framework for understanding how celestial conjunctions could affect human affairs, and was translated into Latin in the eleventh century under the title *On the Great Conjunctions* (*De magnis conjunctionibus*).[45] Defending a doctrine of astral determinism that seemed to contravene the Christian doctrine of free will was fraught and potentially heretical, but *On the Great Conjunctions* was enormously influential on Latin intellectual culture. Several of Bacon's contemporaries also defended the use of Abū Maʿšar's theory of astral influence and determination; for example, Albert the Great argued that celestial influence on bodies was not the same as celestial influence on souls. Bacon held that recent, shoddy translations of Abū Maʿšar introduced errors that were not in the original texts, underscoring again the need to study languages as well as mathematics and other sciences.[46] Certainly, by the 1260s Bacon was focused on texts and authors—especially from Islamdom—that dealt with *experimenta* and the sensorium, putting him out of step with the younger generation of scholars more focused on generating commentaries on the *Sentences* of Peter Lombard, and who took a more critical view of anything that seemed deterministic.[47]

Bacon reflected on this period of independent study and academic freedom, before he became a Franciscan, in the *Third Work* he sent to Pope Clement IV in 1268. "For twenty years I have labored especially in the pursuit of wisdom, abandoning the opinions of the ignorant masses; I have spent more than two thousand pounds on these studies, for books of secrets, various experiments, languages, instruments, tables, and other things; and to seek the friendship of learned people, as well as seeking their close counsel on languages, diagrams, numbers, tables, instruments, and many other things."[48] In this passage Bacon clearly articulated for Clement (and the curia) what the pursuit of knowledge demands: access to books on many subjects, including esoterica; astral tables and charts; scientific instruments; language acquisition; and access to people who know how to read these volumes and use the instruments. It cost Bacon decades and his life's fortune—two thousand pounds, now worth millions today—and required his fellowship among a group of learned, curious, and like-minded people with books, tables, and instruments of their own. Either during this period or shortly after he also completed two works—on creation of mirrors and lenses for incineration and illusion and on the emanation of force—indicative of the new direction of his interests in optics, devices, experiment, and applied science.[49]

One contemporary whom Bacon found exemplary of the potential for his new educational program was Peter (Peregrinus) of Maricourt, who combined familiarity with the university curriculum with applied experience. We know little about Peter; however, he likely studied at the University of Paris in the middle of the thirteenth century and he wrote a treatise on magnetism, *Treatise on the Magnet* (*Epistola de magnete*, 1269), likely while taking part in the Angevin siege of Lucera in southern Italy in 1268–69.[50] Like Bacon, Peter speculated about possible technologies and their applications, but in even greater detail (though in fewer words). He posited the creation of new instruments, such as a new kind of compass needle, a universal astrolabe, and a perpetual motion machine.[51] He also valued an experimental approach to natural knowledge, testing the lodestone (magnetite) in different ways to ascertain and articulate its capabilities. Bacon claimed Peter as an inspiration after working with him in Paris.[52] In the *Third Work*, written perhaps a year before *The Treatise on the Magnet*, Bacon lauded Peter:

> What others, blind, struggle to see, like a bat in the light of the sun at dusk, this man apprehends in the full light of day because he is a master of experiments. He knows about nature through experience, and medicine or alchemy, and all things terrestrial and celestial. Indeed, he would be ashamed if some layman, or a crone, or a soldier, or a rube from the countryside would know something of which he himself was ignorant. He has carefully investigated the smelting of metal ore and how to work gold, and silver and other metals and minerals; he has mastered all sorts of arms used in military service and in hunting, besides which he has carefully investigated all matters relating to agriculture and surveying and the work of agricultural laborers; and he has even closely examined the experiments, incantations, and devices of hags and sorcerers; and likewise

> [examined] all the illusions and devices of conjurers so that nothing that is to be known might escape his notice, and he knows more than enough to condemn magic and all deceitful things.[53]

According to Bacon, investigating the natural world required mastery of traditional university subjects, like astral science, alongside subjects that combine text and practice, like alchemy or medicine. But one must also have real-world experience in subjects that deal with the material world, like metallurgy and mining, as well as applied mathematics, like the military art of ballistics. One must also have the right character and habits of mind to investigate nature: a willingness to test received wisdom; careful, sustained attention; and the discernment and moral fiber to eschew "magic and all deceitful things." Bacon also asserted that people from different registers, with different kinds of experience, could be a source of knowledge for the natural philosopher. Soldiers could offer insight into ballistics; folk remedies and crone wisdom were a source of medical and botanical knowledge; and the rustic who toiled in the fields had experiential knowledge of meteorology, astral observation, and agronomy.

Rustics, hags, sailors, soldiers, and artisans were also rich sources of lore, which could be put to the test by *scientia experimentalis*. To convince his reader Bacon tested several commonly held ideas about the natural world and disproved each of them. The belief that diamonds can only be broken by goat's blood? No, "without that blood one can easily break a diamond. For I have seen this with my own eyes, and this is necessary, because gems cannot be carved except with fragments of this stone."[54] The notion that the beaver, when hunted for its musk glands, castrates itself to save its life? No, "the beaver has these glands under its breast, and both the male and female produce these glands."[55] The belief that the male beaver would mutilate itself to appease its human predator commonly appeared in bestiaries, which drew on material from encyclopedias and travel narratives to depict the natural world in terms of Christian fable and morality (Figure 1). In this example, the moral is that the virtuous Christian (the beaver) must cut off his vices (his testicles) and cast them out and into the devil's (the hunter's) face. The devil will see that the virtuous Christian has nothing for him and so will spare him further. By using this example to indicate what knowledge and certainty *scientia experimentalis* offers, Bacon is indicating that *scientia experimentalis* provides the devout Christian a path to communion with God, a path built on accurate natural knowledge. Another commonly held belief proved false

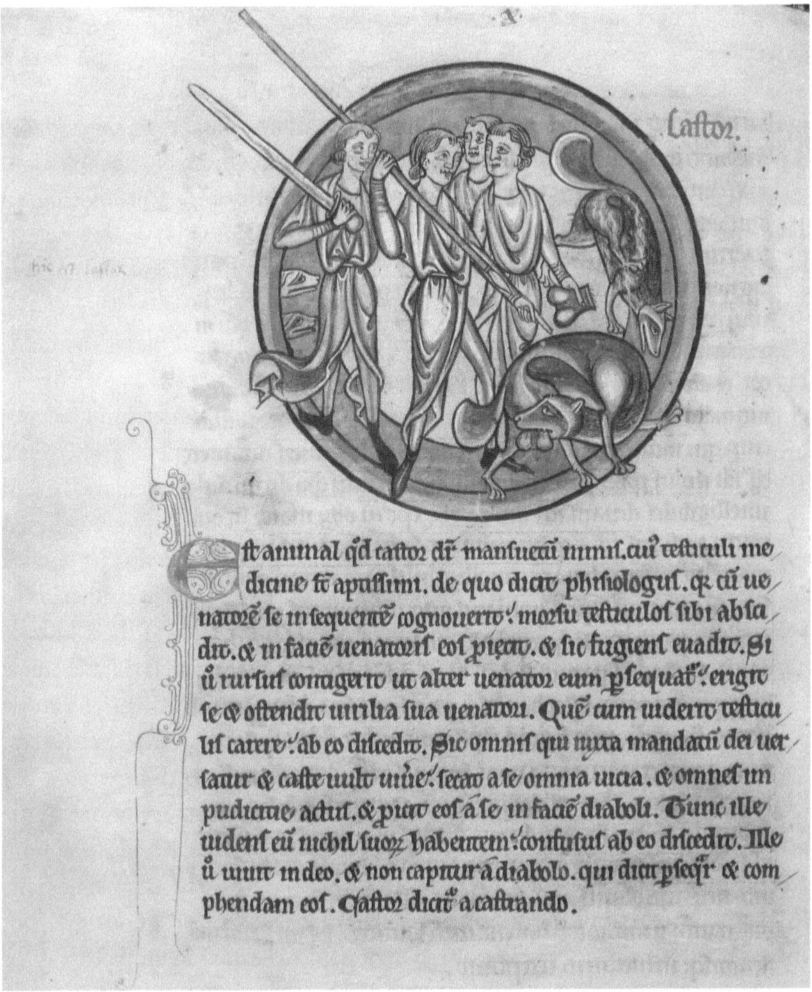

Figure 1. A beaver hunt. One beaver is biting off his testicles when cornered by a group of hunters; another beaver in the scene lacks testicles; one of the hunters has a pair of beaver testicles in his hand. From a bestiary made in England, middle of the thirteenth century. London, British Library, Harley MS 4751, f. 9v.

by trial is that hot water in a container will freeze more quickly than cold water in a container. Even though "it is argued that contrary is excited by contrary, just as when enemies face off against each other. But it is certain that cold water freezes more quickly for anyone who makes the experiment."[56] Observation and experimentation to prove or disprove commonly held beliefs, carried out by a learned and careful experimenter, are vitally important to correct erroneous knowledge and to provide certainty.

When practical experience is valued on par with book knowledge, the ranks of who can be considered authoritative and knowledgeable expand, requiring scholars to engage with a wider range of knowledgeable people than if considering only textual authorities. Yet Bacon was not promoting an egalitarian view of knowledge; his use of "ashamed" in this passage reveals his condescension. Yes, full investigation into nature must draw from people in different registers and walks of life but Bacon nevertheless considered the enterprise of knowledge-production as invested in authority and hierarchy. He also saw it as extractive of the expertise—the experiential knowledge—of people he considered not qualified or fully capable to understand or even investigate nature, people like uneducated soldiers and farmers, and crones. Authoritative texts may be mistaken, due to ignorance of particulars, lack of attention to *experimentum*, or poor translation, but Bacon strongly believed in the divide between the ignorant masses and the wise few.[57] His condescension rests on the belief that he (and Peter of Maricourt) are superior to the rustics and the hags: They know more, and the quality of their knowledge is better. Like the men of whom Aristotle spoke, Bacon and Peter know both the fact and its cause, they have both rational and experiential knowledge. To translate *empeiria* and *experientia* into *scientia* requires the work of a learned and careful experimenter, rather than a scholar or philosopher. Bacon also made clear, here and elsewhere, that another purpose of one's erudition and discernment was to identify and repudiate magic and deception. Effects or phenomena that might seem to be magical—or that magicians might claim to conjure—were not caused by demons or fraudsters, but through a deep understanding of causes joined to skilled manufacture.

More importantly, the moral character of the experimenter was critically important; if one relied on accounts from others (as necessity required), how could one know if the information they provided was credible and accurate unless they were also of the proper moral fiber? Aristotle could attest to more than Ptolemy regarding the regions of the world, because "Aristotle, on the authority of Alexander, sent two thousand men throughout different parts of the world to prove through experience all things that are on the surface of the earth, as Pliny says in *Natural History*."[58] Because of the need to rely on others' observations and experiences, the character of the experimenter was of paramount importance. Aristotle's great wisdom and discernment extended to his ability to choose trustworthy, accurate men to send on this expedition. According to Bacon, virtuous men could make observations and could reason from argument and phenomena, but not uneducated soldiers, farmhands, or women.

Secrets and Powers

Bacon was concerned with the application of natural knowledge that *scientia experimentalis* could foster, and how this knowledge could be used to intervene in the natural order for political, as well as epistemic, gain. His admiration for Peter of Maricourt, who looked to what kind of natural knowledge could be gained and applied outside the classroom or the cloister, reflects this concern. This interest in applied knowledge, in the utility of knowledge for rulership (alongside personal attainment of *sapientia*), and in considering the kind of empirical knowledge from people like soldiers, miners, farmers, and crones, indicates that Bacon's context was the court at least as much as the classroom and the cloister. Bacon drew on the courtly sciences, such as engineering, navigation, and alchemy—those branches of natural knowledge that focus on materials and practical application, where the goal is to intervene in the natural order to improve the quality of human life, or for domination, or both. He looked to Alexander the Great—the student of Aristotle and great empire-builder—for an example of the potential benefits of erudite knowledge that combined knowledge from the sensorium alongside textual knowledge working in the service of political power.[59]

The mirror-for-princes ruling handbook *The Secret of Secrets* (*Secretum secretorum*) captivated Bacon. He first encountered it in the 1240s and spent decades, on and off, studying it, producing his own glossed edition of the work in the 1270s. This text, which is a Latin translation of the Arabic *Kitāb sirr al-asrār*, circulated as a letter that Aristotle sent to his most famous pupil, Alexander the Great. In it, Aristotle imparted to Alexander all the knowledge that he had withheld from his earlier works out of concern that this knowledge could fall into the wrong hands. But Alexander would need this knowledge to rule his empire as a wise and powerful emperor. Translated in full for the first time in the 1220s, the text encouraged Bacon's focus on the utility of natural knowledge, especially in the service of worldly power. The prologue of *The Secret of Secrets* puts the text into the context of an existing and robust epistolary relationship between Alexander and Aristotle, and in many instances *The Secret of Secrets* appears bound together with Latin works on Alexander or vernacular texts on the Alexander tradition.[60] Both *The Secret of Secrets* and the historical literature on Alexander the Great focus on rulership, natural knowledge, and exploration for knowledge-acquisition.[61] Episodes and examples drawn from the life of Alexander and *The Secret of Secrets* appear in a shorter text Bacon wrote to clarify his views on *scientia experimentalis* and the wonders it could implement through legitimate human art. In *Letter on the Hidden Powers of Art and Nature, and on the Invalidity of Magic* (*Epistola de secretis operibus artis et naturae et de nullitate magiae*, 1270s) Bacon described how *scientia experimentalis* certified theoretical natural knowledge and suggested how human ability could exploit natural phenomena in the service of power.

The Secret of Secrets is a doubly experimental text. A practical guide focused on the experience of the ruler; it also contains information pertaining to both knowledge gained through theory and from experience. Although *The Secret of Secrets* circulated as a genuine Aristotelian text, it is pseudo-Aristotelian. Its origins are shrouded in doubt, although it is possible that it may have been based on Roman spuria.[62] A loose collection of pseudo-Aristotelian texts seems to have been translated, via Syriac, from Greek into Arabic during the early eighth century under the Umayyads.[63] It was then recompiled and organized into a princely handbook, *The Book of the Science of Governance, on the Proper Ordering of Statecraft* (*Kitāb al-siyāsa fī tadbīr al-riyāsa*) in the later ninth or early tenth century by (pseudo-) Yahyā ibn al-Bitrīq (fl. 815 C.E. / 199 A.H.), during a period in which much of the Greek corpus on medicine, natural philosophy, and ethics was translated into Arabic.[64] The text also is exemplary of *adab* literature—manuals for deportment and self-improvement, that included *compendia* of natural philosophy, physiology, cosmology, divination, poetry, and moral philosophy.[65] Over the next few centuries, some material was added and other material reorganized, and the text became more commonly known by its subtitle, *Kitāb sirr al-asrār* (*Book of the Secret of Secrets*) (Figure 2).

The text was partially translated into Latin and called *Secretum secretorum* by John of Seville around 1125, and it circulated piecemeal throughout Latin Christendom for the next century, until largely superseded by the complete translation (of a longer version of the Arabic text) by Philip of Tripoli around 1220.[66] *The Secret of Secrets* circulated within the papal and imperial courts from the early 1230s. Frederick II's court was in close contact with papal courts, as well as crusader states and political centers throughout Asia Minor and North Africa, and through at least one of these avenues Michael

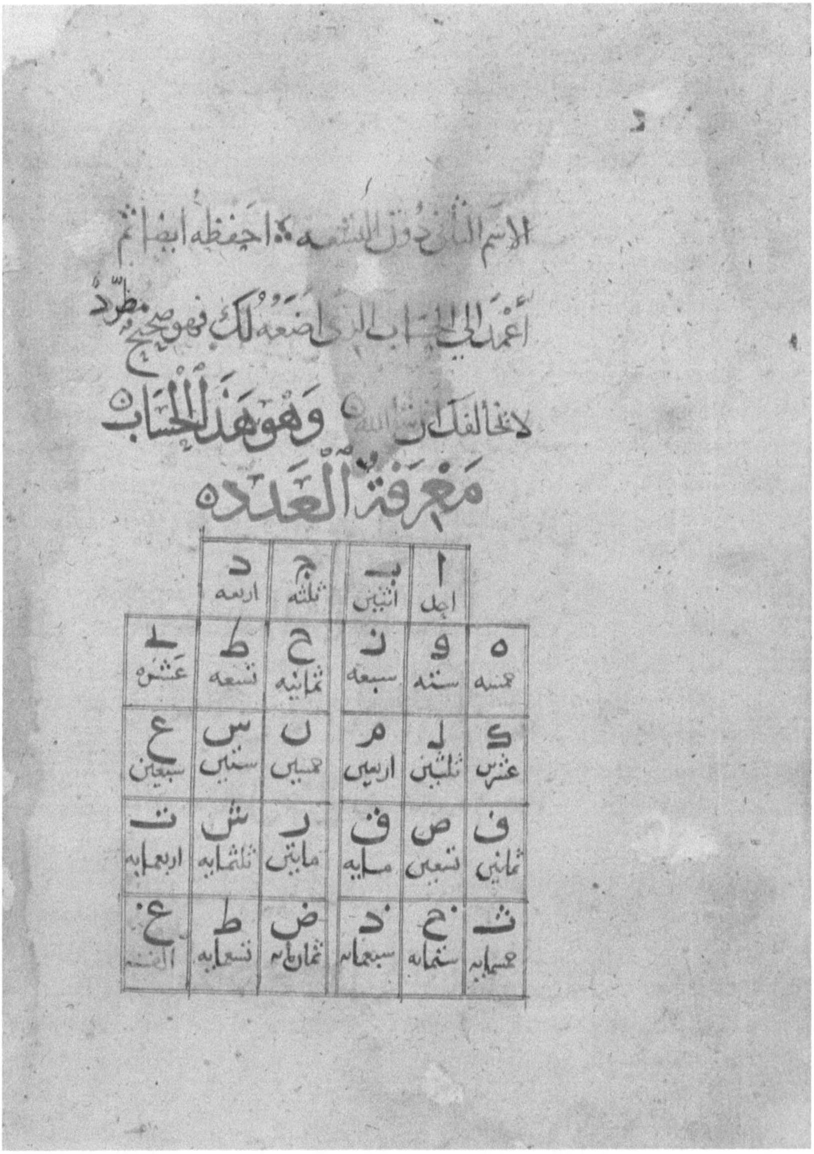

Figure 2. *Kitāb sirr al-asrār* circulated extensively in Arabic, including finding audiences at the highest levels of the court. This copy was likely made in Mosul for the ruler Nur ad-Din Arslan Shah (ca. 1200 CE / ca. 570 AH), who is named on the first folio. This table gives the numeric values for different letters of the alphabet, for numerology, a branch of the divinatory sciences. Philadelphia, Pennsylvania, University of Pennsylvania Library, MS LJS 459, f. 114v.

Scot became familiar with the work in the 1230s.[67] As previously discussed, both Scot and Bacon spent time in Paris as scholars. Although it seems doubtful that they overlapped there (Scot likely died before Bacon arrived), Bacon encountered *The Secret of Secrets* in Paris in the late 1230s, by which time the text included information on complexion, health and disease, geography, cosmology, and moral philosophy, as well as knowledge of particulars, such as geomancy, talismans, the powers of animals, and alchemy.[68]

Bacon prepared an edition of *The Secret of Secrets* in the 1270s, complete with diagrams and emendations (Figures 3–8). In it, he noted that he had consulted four copies in Oxford, but had seen other, "perfect" copies back in Paris, likely in the 1230s and 1240s.[69] Bacon referenced *The Secret of Secrets* as an authoritative Aristotelian text in his own writings from the 1260s and 1270s, during a period of disagreements among the intellectual community in Paris and in the Franciscan Order about doctrinal orthodoxy in Aristotelian natural philosophy.[70] Bacon argued that although the possibilities in the *Secret of Secrets* might seem to contravene Christian doctrine or canon law, faulty translation was the cause of the misunderstanding. "Concerning these natural sciences, which may be improperly called geomancy, hydromancy, aeromancy, pyromancy, which are really parts of philosophy, as Aristotle asserts in this book, but the translator did not have proper names for these sciences in Latin, so he took the names of magical sciences which are similar to some true sciences."[71] But it was pointless to attribute to magic those effects that had a natural cause, and it was also a mark of ignorance. The truly learned understood the causes of wonderful phenomena and had no need of magical explanations.

Both Scot and Bacon, with interests in text-based natural philosophy and *experientia*, took up the framework that *The Secret of Secrets*—believed to be an authoritative Aristotelian text—provided for relations between the ruler and his wise counselor. In his commentary on the text, Bacon made clear the stakes of the kind of knowledge *The Secret of Secrets* offered. At one point, Alexander asked Aristotle if he should wipe out an entire people, whom he considered evil and corrupt. No, replied Aristotle, you should use a talisman—a physical object with an inscription that connected the object to the power of a celestial body—to change their complexion—their biology and temperament—to one that was more pacific and easily ruled.[72] Bacon also referenced the doctrine of conjunctionism and celestial influence throughout his commentary: for example, urging physicians to bring their medicines outside to absorb the beneficial cosmic rays from particular conjunctions.[73]

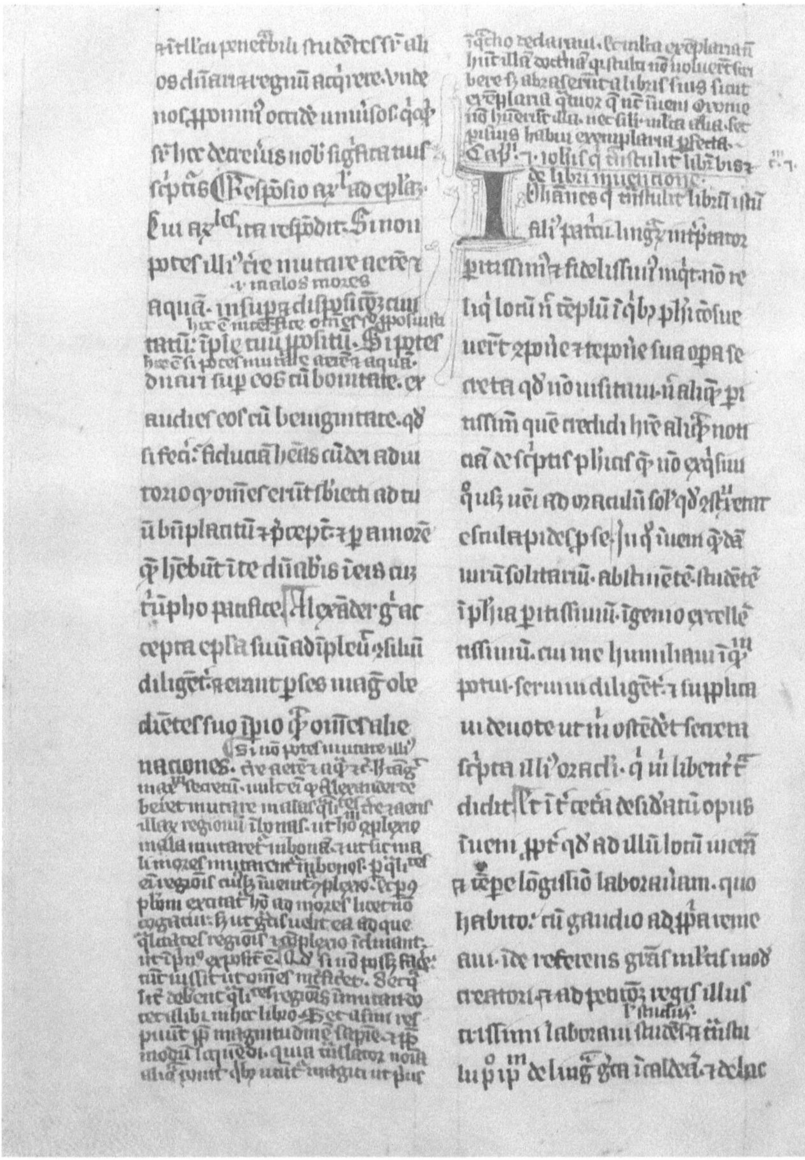

Figure 3. Text of *The Secret of Secrets* with Bacon's comments in smaller text alongside. Oxford, Bodleian Library, MS Tanner 116, f. 19v.

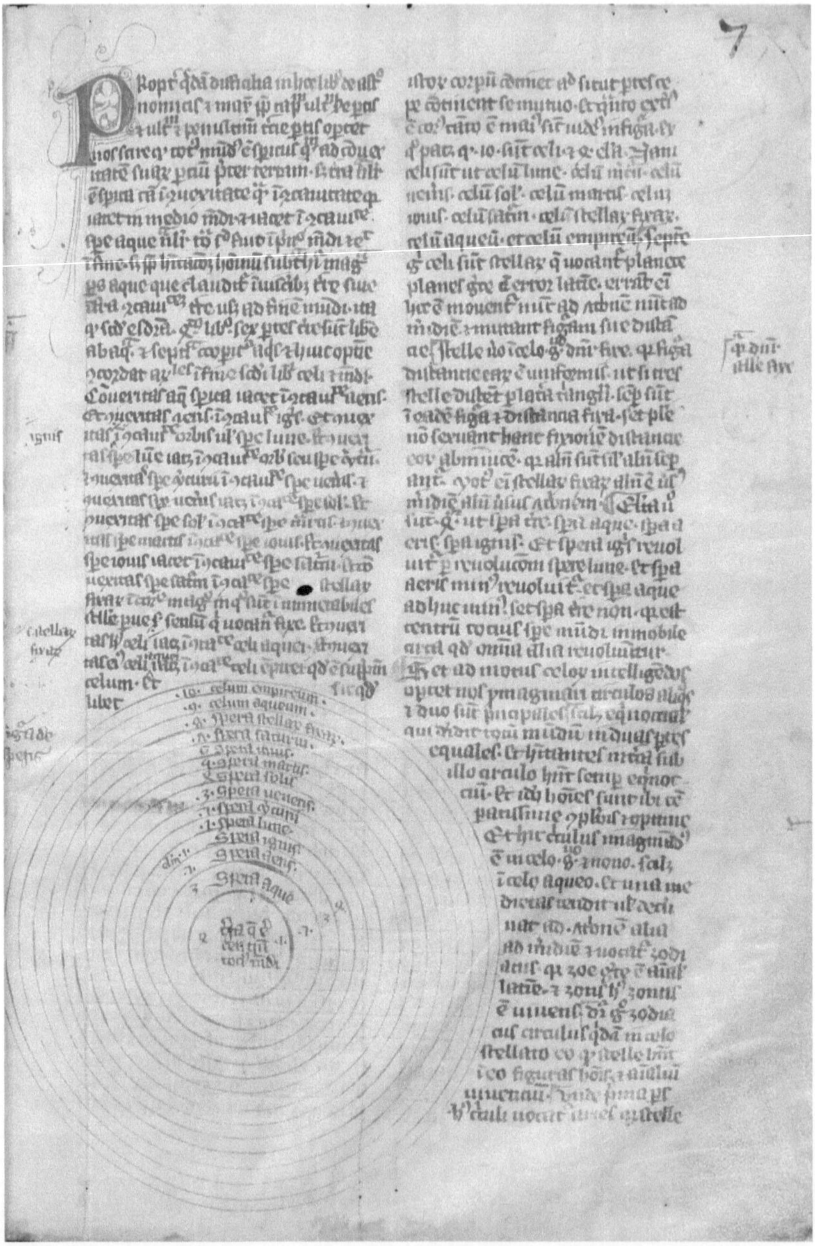

Figure 4. Diagram of the cosmos, following the geocentric Ptolemaic model. Oxford, Bodleian Library, MS Tanner 116, f. 6r.

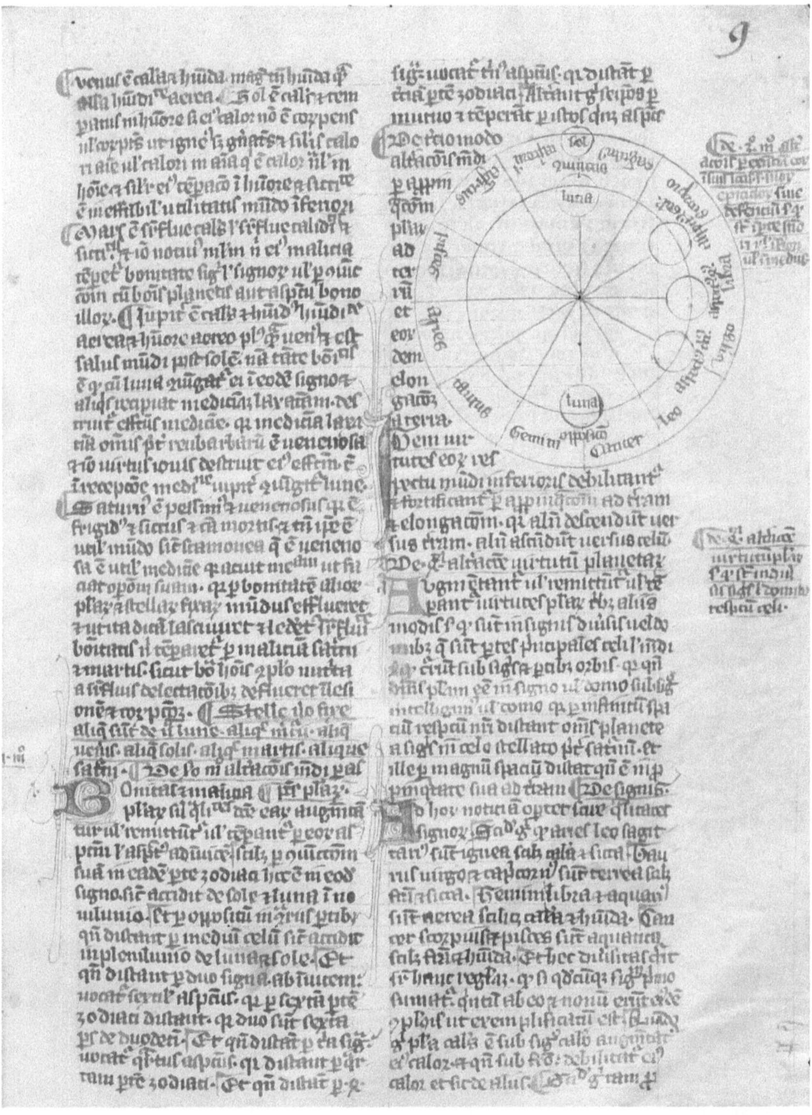

Figure 5. Zodiac diagram of the moon in different aspects. Oxford, Bodleian Library, MS Tanner 116, f. 7r.

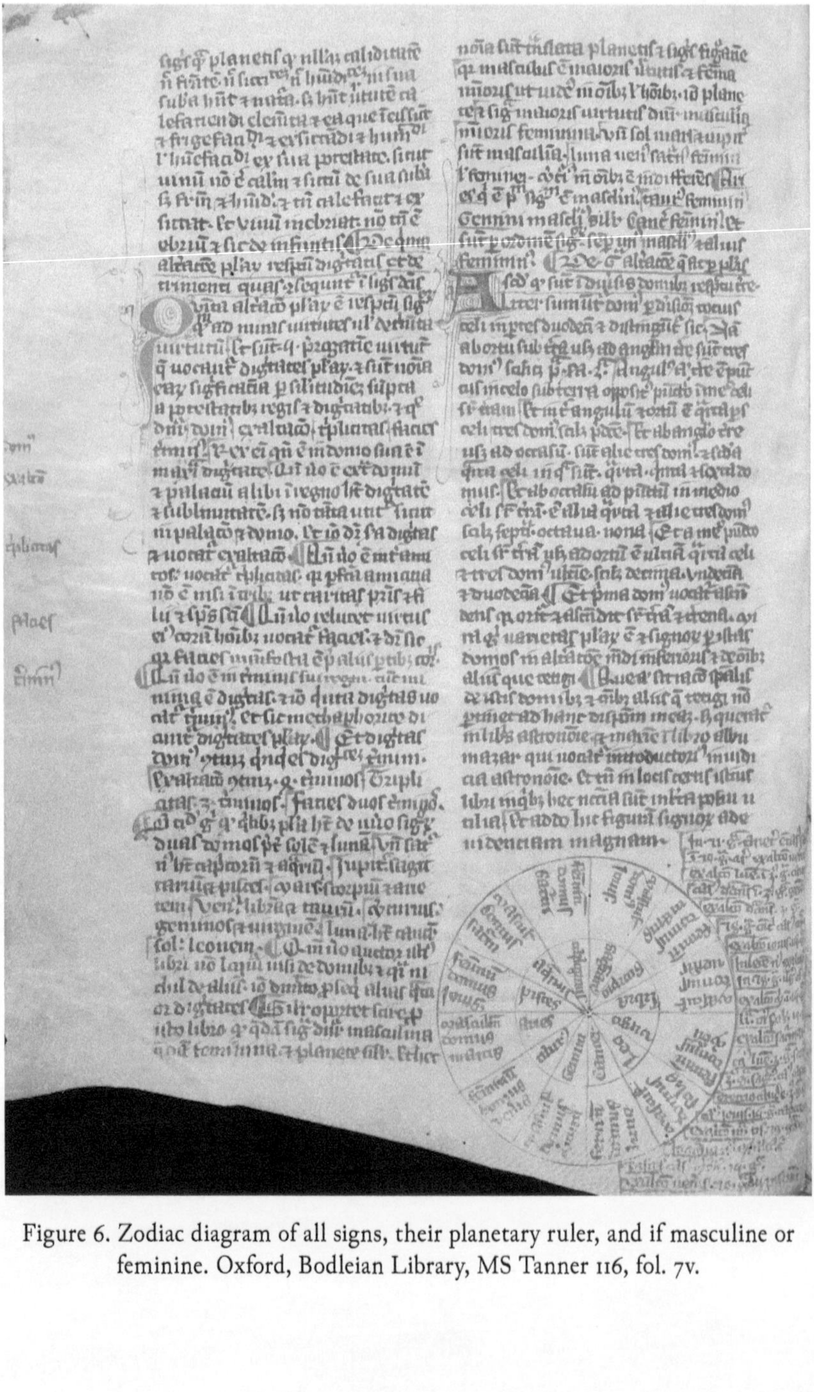

Figure 6. Zodiac diagram of all signs, their planetary ruler, and if masculine or feminine. Oxford, Bodleian Library, MS Tanner 116, fol. 7v.

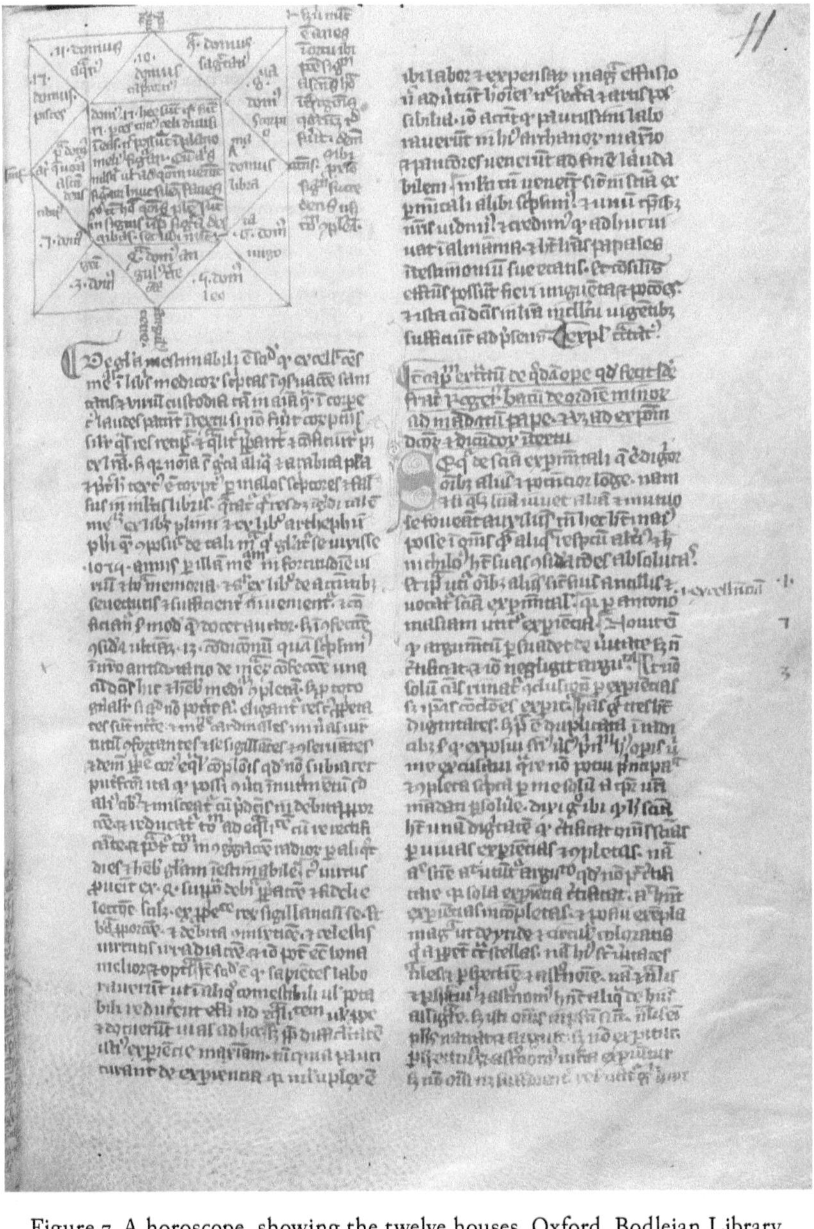

Figure 7. A horoscope, showing the twelve houses. Oxford, Bodleian Library, MS Tanner 116, fol. 8r.

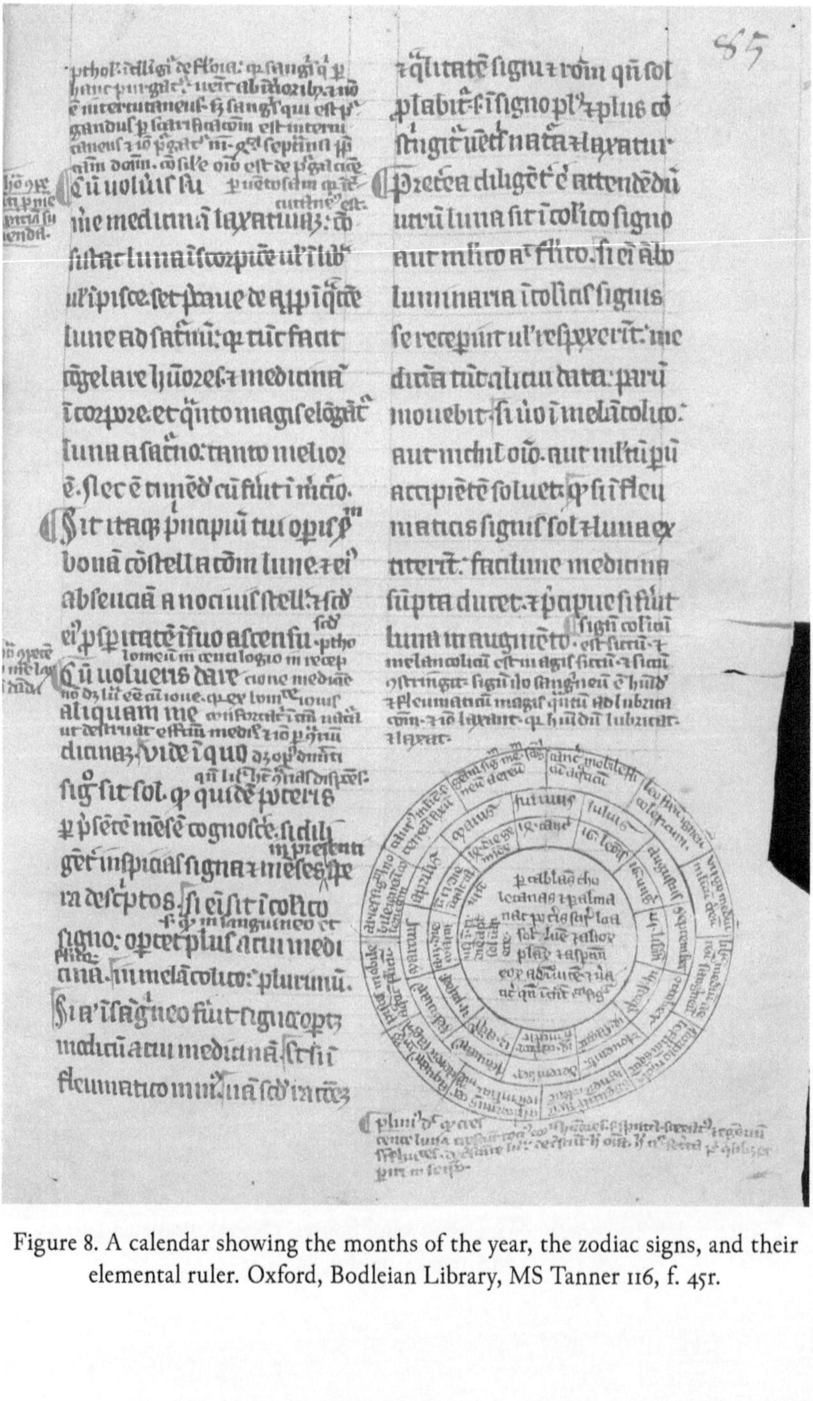

Figure 8. A calendar showing the months of the year, the zodiac signs, and their elemental ruler. Oxford, Bodleian Library, MS Tanner 116, f. 45r.

Bacon, convinced of the utility of *scientia experimentalis* and the kind of knowledge that texts like *The Secret of Secrets* could bestow, prepared his edition of this mirror-for-princes for an unnamed prince (possibly Henry III or Edward I of England) addressed throughout the text as "you."[74]

During this same period, Bacon completed his short text on the potential of *scientia experimentalis* and the natural and technological marvels it could enable, *Letter on the Hidden Powers of Art and Nature, and on the Invalidity of Magic*. We know little about the circumstances of its composition or its intended audience. It was likely completed while Bacon was in Oxford, and it seems, in its tone, focus, and level of detail, to be directed to a ruler or courtier, rather than to a natural philosopher. However, like Bacon's edition of *The Secret of Secrets* it was also completed during a period of doctrinal upheaval within the Order, and it is possible that Bacon wrote *Hidden Powers* to clarify the difference between *scientia experimentalis* and magic, which he considered fraudulent and pointless. The first line of the extant text suggests only that Bacon was responding to a request for information.[75] As Bacon laid out in *Hidden Powers*, not only could *scientia experimentalis* help discern the deceptions practiced by magicians and charlatans, but when yoked to skilled human artifice it also made possible incredible inventions and processes, which could in turn be used to strengthen Christendom, or a Christian realm. The combination of learned and experiential knowledge in the service of the ruling elite—courtly science—is the entire point of *Hidden Powers*, just as it is the entire point of *The Secret of Secrets*.

As is common with Bacon's work, *Hidden Powers* repeats several of the points made elsewhere, especially in the fifth and sixth books of the *Great Work*, on optics and *scientia experimentalis*, respectively. The fifth chapter of *Hidden Powers*, on optical illusions and lenses, is titled "Artificial Optical Experiments," recalling Ibn al-Haytham's articulation of the importance of using lenses and other visual instruments to measure and investigate optical phenomena. Lenses and other practical knowledge of optics could also be used for reading, measurement, and espionage. "For glasses can be so shaped that things placed at the greatest distances appear most near and conversely; so that we could read the most minute letters at an incredible distance, count things however small."[76] Bacon had seen lenses that could be used this way, though perhaps not to the extent that he described here. Lenses similar to the ones he described existed in this period (the middle of the thirteenth century) and it is likely that eyeglasses with lenses like these were in use within twenty years of Bacon writing *The Great Work* and *Hidden Powers*.[77] And

Bacon had written a short treatise on burning mirrors and in his *Third Work* he wrote that he had dispatched the work along with his young student or protégé John, and a spherical crystal lens to do experiments with burning lenses.[78] The following chapter is "On Marvelous Experiments," on natural marvels such as the magnet and Greek fire, making evident the conceptual association between *experimenta* and natural particulars.[79]

Bacon also included new material in *Hidden Powers*. He expanded on the possibilities *scientia experimentalis* furnished to explore new territory and consolidate power as part of the second prerogative of *scientia experimentalis*: to make instruments or machines using natural laws and powers. "First, those things achieved through the design and reckoning of skill alone: Now an instrument for sailing without oarsmen can be made such that the largest ships, both riverboats and seagoing vessels, can be moved under the direction of a single man at a greater speed than if they were filled with men. And a chariot can be made that moves at an unimaginable speed without animals; such we think to have been the scythe-bearing chariots with which men fought in ancient times."[80] Other inventions could enable new networks, or link previously separated populations, "such as bridges that span rivers without pillars or any support, and machines and unheard-of inventions."[81] Others could be used for civil engineering and civil defense: "And an instrument small in size for raising and lowering almost infinite weights; at times there is nothing more useful than this, for, by an instrument three fingers high and just as wide, or even smaller, a man could snatch up himself and his friends and raise and lower them from every danger of prison."[82] Devices could be made using magnetism that would render entire populations powerless: "[An] instrument could easily be made by which one man could drag a thousand men against their will toward himself, and attract other things in the same way."[83] Furthermore, the use of lenses and knowledge of perspective could reduce an army or town to ashes, or terrify it with illusions. Knowledge of lenses and mirrors could be used for spying, "so that hidden things appear evident."[84] This emphasis on military utility echoes Bacon's earlier assertions in his works for Clement IV that *scientia experimentalis* could help defend Christendom from her enemies.

Bacon's edition of *The Secret of Secrets* and his completion of *Hidden Powers* date from the 1270s, and both contain material and ideas also found in Latin and vernacular literature on Alexander the Great. Just as the former is addressed to Alexander the Great, throughout the latter Bacon is casting himself as Aristotle to some latter-day Alexander. During the long medieval period, Alexander the Great was one of the few exemplars for good rulership from the ancient world. Although there are many variations on the legends of Alexander, one thing they share are Alexander's defining characteristics: his interest in exploration, in first-hand experience that went beyond ordinary human knowledge, and his ingenuity.[85] In Latin and Old French *ingenium* and *engin*, respectively, are the terms that refer to this kind of inventive spirit. But these terms and their cognates also encompass other meanings, such as intellectual wit, chicanery, deception, stratagem, and skilled technical knowledge.[86] The Anglo-Norman *Roman de Toute Chevalerie* (ca. 1180) was composed at the court of Henry II Plantagenet (1133–89) and based in part on a text that circulated widely as a letter from Alexander to Aristotle about the marvels he encountered on his travels, the *Letter of Alexander the Great to Aristotle* (*Epistola Alexandri Magni ad Aristotelem*). Like Alexander, Henry had a reputation for dynamism and military victory, created a new empire through conquest, and had been educated in philosophy and military arts. The author, Thomas de Kent, ends his introduction to Alexander with this summation: "He was brave and victorious, wise and ingenious."[87] In the contemporary *Roman d'Alexandre* (ca. 1185) by Alexander of Paris, Alexander's extensive knowledge of the natural world and his technological ingenuity are two of his essential qualities.[88] And in many Alexander-texts from this period his endeavors are discussed in terms of experience and proof: *experimentum* and *esprover*. In multiple versions of his biography he devises a flying machine, powered by griffins, and a glass diving bell to explore rivers and oceans.[89] Both inventions enable Alexander to gain new knowledge through experience. His devices allow him to explore nature's secrets beyond what was known at that time. Perched on his flying machine he could experience the world as no human before him. From the safety of his diving bell he could observe the creatures at the bottom of the ocean. Moreover, he employed his knowledge for conquest, as well as exploration. In *Roman d'Alexandre*, Alexander's successful campaigns against several cities hinge on his inventions, such as the floating siege towers that allowed him to capture the city of Tyre.[90]

Alexander's rule is enlightened by his traditional education, first-hand experience, and technical knowledge. His education from Aristotle, his tutor and interlocutor, included philosophy, astral science, and geography, as well as the secrets of nature.[91] In the *History of Battles* (*Historia de preliis*), the tenth-century Latin prose retelling of the late antique Greek version by Pseudo-Callisthenes, Alexander designed a flying contraption that used winged animals to see the earth in a new way. "I planned with my friends that I should build a contraption, so that I might ascend into the sky and see what may be seen from the sky. I designed and built the device, where I would sit, and caught griffins and bound them with chains and I put a pole at the end of which was food in front of them, and they began to take off into the sky."[92] In this account, Alexander appears as an active, agentic ruler, designing and manufacturing the device that will let him "see what may be seen." His allegiance to autoptic knowledge, his ability to innovate, and his interest in discovery are all marks of the ideal ruler. The *Roman d'Alexandre* explicitly links experiential knowledge to the confirmation of text-based learning, via measurement. "Alexander forged the path up to the sky when his golden chair, attached to four griffins, was carried up; and his thought [having been] enlightened by astral science to the extent that that he knew the compass of all the stars."[93] Speculative, imaginary, or legendary machines and devices appear frequently in literary texts in romances in the twelfth and thirteenth centuries, and literary texts were conceived of as spaces to stage thought experiments, and to engage with different forms of experience.[94]

The theme of Alexander's capacious education and interest in the world is evident in the pictorial representations of Alexander and his education from this period. In the image from a copy of *The Secret of Secrets* made for King Edward III, Aristotle teaches Alexander about the effects of celestial bodies on earthly things; his gestures conform to standard figural representation of teaching in this period (Figure 9). In a slightly later copy of the French translation of *History of Battles* (*Historia preliis*), titled *The True History of the Good King Alexander* (*Le Livre et le vraye hystoire du bon roy Alixandre*), in one half of a vignette Alexander receives instruction from Aristotle in the traditional setting of the schoolroom—including books—and in the other he is outside, receiving education from different teachers (Figure 10). The Old French version was widely copied (sixteen manuscripts are extant, the earliest from 1290) and illuminated with a consistent picture-cycle.[95] In two images from an early fourteenth-century copy, Alexander appears unmistakably kingly—seated in majesty with crown and scepter—as he explores the earth from up in the air to under the sea. In the first image the importance of experience is evident from the expressions of the rapt eyewitnesses to Alexander's flight (Figure 11). In the second image, the sea teems with creatures, including two individuals who point and gesture toward Alexander in his glass barrel—they can't believe their eyes (Figure 12). The image vividly depicts

Figure 9. Alexander receives instruction from Aristotle on the influences of the planets on plants. *Secretum secretorum*, 1327. London, British Library, MS Add. 47680, f. 49.

Figure 10. Alexander studies with Aristotle inside the classroom, and outside. *Le Livre et le vraye hystoire du bon roy Alixandre*, 1420, London, BL MS Royal 20 B XX, f. 10v.

the stakes of experiential knowledge—the ability to see things that were previously unknown or lived only in the imagination.

Bacon was clearly familiar with these legends, as he cited a few of them in *Hidden Powers*. He insisted that "an instrument for flying can be made, such that a man sits in the middle of it, turning some sort of engine by which artificially constructed wings beat the air in the way a flying bird does."[96] Alexander's diving bell also appears in *Hidden Powers*. "And instruments can be made for walking in seas and rivers, right down to the bottom, without bodily danger. . . . For Alexander the Great used these to see the secrets of the sea, according to what Ethicus the astronomer says."[97] Bacon's *Hidden Powers*, the *Secret of Secrets*, and multiple versions of the medieval Alexander-legend all express the idea that Alexander's reign was a dynamic blend of inventiveness and utility. This past example offers Bacon a model for the potential future that he conjured for his audience.

As Bacon set forth, *scientia experimentalis* was the key to regaining past knowledge and to acquiring new knowledge. *Scientia experimentalis* enabled the combination of past knowledge and present knowledge to envision a future Christendom that was morally and politically indomitable. The impossibly fast chariots that Bacon mentions are the scythe-bearing chariots "of ancient times." Optical devices that allow one to see things that are very

Figure 11. Alexander in his flying machine. The text reads: "Comment Alixandres se fist porter en l'air as oyseals que l'en apele grif." *L'Ystoire du bon roi Alexandre*, 1320. Berlin, Kupferstichkabinett, Staatliche Museum Preußischer Kulturbesitz, MS 78 C 1, f. 66r.

Figure 12. Alexander in the diving bell. The text reads: "Comment Alixandres se fist avaler en la mer ou tonnel de voirre." Berlin, Kupferstichkabinett, Staatliche Museum Preußischer Kulturbesitz MS 78 C 1, f. 67r.

far away with perfect clarity are also inventions of the past and were used to subdue Britain. "Thus, it is thought Julius Caesar on the shores of Gaul discovered by huge mirrors the disposition and locations of the camps and cities of Great Britain."[98] This is a neat inversion, as Bacon argued in this text for the importance of *scientia experimentalis* and the inventions it affords to strengthen Britain. Even the most extraordinary device, the magnetic armillary sphere that turns with the daily motion of the heavens, is based on an invention of Ptolemy, according to Bacon. These objects highlight Bacon's point about the need for experience to work in tandem with textual knowledge. The moving armillary sphere, submarine, optical devices, and other machines are ones that Bacon and his contemporaries only know through texts, which is why he could not describe how to make them. He employed the common trope of using familiar items to describe unfamiliar technology, and summoned a future in which familiar objects, like ships or chariots, would become new.[99] This temporal equivocation also echoes the temporal plurality found in contemporary histories, written in vernacular, or *romanz*. In these romances, stories from antiquity, like the Trojan War, or King Arthur's court, appear refashioned and updated for courtly audiences in a recognizable contemporary setting.[100] Alexander's exploits in the distant past yielded important experimental knowledge—of plants, places, peoples—and devices for gathering experiential knowledge, as conveyed in different versions of his travels. Bacon wrote from a position of loss, as these inventions and instruments used to exist, but did not at the time of Bacon's present. Yet this knowledge, in part, provided the foundation for his vision of the future, as the accounts of past devices prove that their presence in the future is not theoretical or speculative.

Bacon used a wide array of sources to invoke the future. *Cojointure*, the skilled combination of multiple sources that yields a new invention in literary turns, here yokes past, present, and future together. By going outside of the university context and using source material that circulated in courtly contexts, such as *The Secret of Secrets*, Bacon was able to approach the past and present as dialectical forces that could combine to create a brilliant future. His insistence on the ancient precedents for some of these inventions, along with his frequent invocation of ancient sources (Ethicus, Ptolemy, Aristotle, Pliny), makes it clear that, far from rejecting claims to authority in his work, as later historiography claimed, Bacon relied on them.[101] His inventions are Janus-faced, oscillating between the past and the future, and between the familiar terrain of accepted authorities like Ptolemy and the newer, improved ideas of foreign philosophers like Ibn al-Haytham. Bacon wrote from a temporal perspective between two poles. Instruments, machines, and devices once existed, do not exist now, but could be recovered in the future. Other inventions, unheard of, are rendered conceivable in terms of familiar technology—ships unlike any other ships, self-propelled chariots—that will allow the future to exceed both present and past.[102] The future might resemble the past but would also go beyond the rediscovery of ancient technology. "An almost infinite variety of things can be made . . . instruments and unheard-of engines."[103] Bacon's invocation of past and future knowledge from a present perspective of relative ignorance echoes his account of the third purpose of *scientia experimentalis*—it provides access to past and future knowledge. Bacon combined *renovatio* with *innovatio*: the recovery of past knowledge, the adoption of more recent knowledge, and a progressive future that goes beyond the past and present. Indeed, the possibilities seem endless to one who understands the full potential of nature. "After I have examined things of this sort, I find nothing difficult to believe, when I carefully inspect it, neither among divine things nor among human things."[104] Bacon articulated a regime of historicity that is progressive, but not linear.

Bacon drew on multiple sources for his concept of *scientia experimentalis* and the kinds of devices and processes that it facilitated. He read newly translated Latin versions of Arabic texts in optics, secrets, and astral science alongside contemporary treatises on natural particulars and *experimentum* and ancient history to articulate the possibilities of *scientia experimentalis*. He drew from the academic register of natural philosophy and the literary register to articulate the possibilities of human art yoked to nature. The review of his influences demonstrates the widespread interest in knowledge through sense experience. Far from being unusual in his interest in experimental thinking, Bacon was part of a lively intellectual current characterized by interest in empiricism, testing, and applied knowledge. Bacon combined his influences with his own ideas about the purpose of technology to enable new knowledge, improve the human condition, and be used in the exercise of power, especially through the creation of devices or inventions. According to Bacon, *scientia experimentalis* complements Aristotelian logic and balances the tools of *argumentum* with the tools of *experimentum* and provides a path to *sapientia*. Additionally, *scientia experimentalis* was vital to political success, and to ensuring a morally, politically, and militarily robust Christendom.

PART II

Chapter 3

Roger Bacon and the Middle Ages

'Twas the *Popes* smoak which made the eyes of that Age so sore, as they could not discern any open hearted and clear-headed soul from an *heretical Phantasme*. The silly Fryers envying his too prying head, by their craft had almost got it off [Bacon's] shoulders.

—Preface, *Frier Bacon, His Discovery of the Miracles of Art, Nature, and Magick, Faithfully Translated out of Dr. Dee's Own Copy, by T. M. and Never Before in English*, 1659

Bacon's concept of *scientia experimentalis*, the inventions and devices it promised, and its application as a tool of epistemic certainty and political power all arose from his wider intellectual context, his deep desire for *sapientia*, and his concern over the political and moral crises facing Latin Christendom. Though Bacon wrote widely on many subjects until his death in the 1290s, including theology, his texts on *scientia experimentalis* and subjects related to it, such as optics, garnered the widest circulation and readership in the centuries after his death. Sections from the *Great Work* on optics and *scientia experimentalis* circulated as standalone works. These, along with copies of *Hidden Powers* and alchemical works attributed to Bacon, helped to shape his reputation in later centuries as an alchemist and experimentalist. This reputation dovetailed (and in some cases reinforced) different versions of "Roger Bacon" that emerged also in the early modern period: the necromancer and mage; the father of English science; the prophet of modern science. *Hidden Powers* kindled the imagination of many, as the devices that Bacon described gave rise to further legends about him. These legends also convey

Bacon's posthumous reputation for an interest in experiment, devices, natural philosophy, and the more intellectually and morally questionable areas of divination, conjuring, and necromancy. Moreover, the changing reputation of Roger Bacon (or "Roger Bacon") as a natural philosopher occurred in tandem with a change in the status of Arabic authors, including the ones on whom Bacon had relied. By the end of the seventeenth century Arabic authors and their works had largely been Orientalized within European intellectual culture and were no longer seen as authoritative; later print editions of Bacon's work feature fewer mentions of Avicenna (Ibn Sīna), Alhazen (Alhacen; Ibn al-Haytham), and Alkindus (al-Kindī) than earlier manuscript copies. Though emblematic of larger intellectual currents, this erasure of Arabic authors from Bacon's work served to sever Bacon from his context and made it possible to construct a narrative of his exceptionalism.

The divergent and shifting inventions of Roger Bacon emerged alongside the view that the times in which both Roger Bacon and "Roger Bacon" lived were part of the same period—a long intervening middle period between antiquity and "now." From its earliest appearance this long middle period has been defined in opposition to both Greco-Roman antiquity and to "now." The establishment of a new historical era that ushered in the now of modernity—the Renaissance—coincided with the invention of the medieval period as a necessary, and necessarily inferior, contrast. What began as an aesthetic judgement about the arts and culture in a handful of Italian cities among a handful of writers was adopted into historiography during the Reformation and Counter-Reformation and entrenched during the Enlightenment as the regime of periodization that we still recognize today: ancient, medieval, and modern.

Fame Among the Franciscans

During Bacon's later years and in the decades after his death in the 1290s we see evidence of the impact of his ideas among learned circles relating to the Franciscans and the papal curia. Within the curia, Bacon's writing on optics and visual perception seems to have gained the most traction. This is due in part to the increased interest in optics as adjunct to more widespread interest, among learned circles, in experience and experiment. Optics, as the science of visual perception, was important as a way of understanding experience—knowledge through the sensorium. And because optics relied on demonstration, including instruments and observation, it was important conceptually and practically to *experimentum*. But for Bacon, and many of the Franciscans who took up his work on visual perception, optics was more important for its contribution to metaphysics. Physical vision provided a metaphor to understand spiritual vision—the perception of the divine. Bacon explored the parallels between the two types of vision at length, especially in Book 5 (on optics) of the *Great Work*. For example, Bacon likened the three types of physical vision—direct, refracted, and reflected—to metaphysics. "[And] as to corporeal vision, so also to spiritual vision. And here we can draw multiple comparisons. Direct vision we attribute to God; departure from rectitude by refraction, which produces weaker vision, is suited to angelic nature; reflected vision, which is weaker [still], can be assigned to humans."[1] Bacon sent off the *Great Work*, the *Lesser Work*, and the *Third Work*, and the other works requested by Clement IV in batches between late 1267 and 1268. Although we do not know if Clement himself read these works before he died in November 1268, we do know that Bacon's ideas were taken up by scholars in the curia. Polish natural philosopher Witelo (ca. 1230–ca. 1280s), who was associated with the papal court throughout the 1270s, engaged with the material in Bacon's works as well as with the Latin translation of Ibn al-Haytham's work, *De aspectibus*, in his own treatise on optics, *Perspectiva* (completed ca. 1275).[2] Moreover, given both Witelo's association with Aristotelian translator and natural philosopher William of Moerbeke (ca. 1215–86), and William's place within the curia, it seems likely that William also read Bacon's works; indeed, this may have been what stimulated William's interest in optics.[3] As Bacon wrote many of his most significant works after joining the Franciscan Order, some of his ideas were taken up by his fellow Franciscans during his lifetime and shortly after. Matthew of Aquasparta (1240–1302) relied on the latter's work on optics and the emanation of force, or *species*, when teaching in Paris in the 1260s.[4] John Pecham (or Peckham, ca. 1230–92), later both Provincial Minister of the Franciscan Order in England and Archbishop of Canterbury, likely overlapped with Bacon when both were resident at the Franciscan convent in Paris in the 1260s. Pecham was also part of the curia in the late 1270s, and had access to Bacon's complete works on optics and *scientia experimentalis*. Pecham's own interests in observation, measurement, and sensory knowledge in the areas of astral science and optics bear evidence of a strong Baconian influence.[5]

By the fourteenth century Bacon had a reputation among the Franciscans as a brother devoted to knowledge gained through observation, especially in the science of *perspectiva*, and to the creation of useful marvels. According to one late fourteenth-century Franciscan, Peter of Trau (fl. 1385), Bacon had pursued true knowledge through experiment, rather than merely through writing. He "studied real philosophy rather than merely writing of marvelous experiments. For he had such a subtlety in natural philosophy which he pursued by means of his wonderful demonstrations (of which there is no truer knowledge) than by writing."[6] According to Peter, Bacon had used his knowledge to build a bridge of thickened air across the Channel and had conveyed his retinue to England this way. And he had written of many marvels "and many similar and greater ones in his letter to Pope Clement IV." The account by Peter also mentions that Bacon had pursued his experimental research while at Oxford to the detriment of teaching. He was "more interested in carrying out his wonderful experiments (than which there is no truer knowledge) than writing with a pen or teaching with words," and had used his consummate skill and his boundless wisdom to create extraordinary devices.

> He was so complete a master of optics that from love of experiments he neglected teaching and writing and made two mirrors in the University of Oxford: by one of them you could light a candle at any hour, day or night: in the other you could see what people were doing in any part of the world. By experimenting with the first, students spent more time in lighting candles than in studying books; and seeing in the second their relations dying or ill or otherwise in trouble, they "went down" [i.e., leaving the university] to the ruin of the university: so, by common counsel of the University both mirrors were broken.

In this account, from about a century after his death, Bacon is focused on pursuing his experimental program, putting him in conflict with university administrators (evergreen). According to Peter of Trau he privileged experimental knowledge over textual knowledge and was "more interested in . . . his wonderful experiments . . . than writing with a pen." The capabilities of the first mirror invited Bacon's students to neglect their studies with experimentation, while the second mirror made it possible to see events or phenomena at a great distance. Both inventions in this narrative recall the devices that Bacon discussed in *Hidden Powers*, for both illumination and apprehending events at a great distance. Furthermore, this account, written a century after *Hidden Powers*, recalls implicitly the temporal equivocation of Bacon's earlier text: The previous century was a time of extraordinary invention. Peter's chronicle also links Bacon to the University of Oxford, where he began his studies, rather than to the University of Paris, where he had been both student and master, implicitly emphasizing his Englishness in ways that would be taken up by later writers.

The more well-known legend about Bacon also comes from the fourteenth-century Franciscans and echoes the theme of Bacon's clash with institutional authorities: that he was condemned by his order in 1277 for unorthodox thinking and imprisoned. "The minister general, after consultation with many brothers, condemned and reproved the doctrine of Brother Roger Bacon of England, a Master of Sacred Theology, because it contained some suspected novelties. Because of these novelties the same Roger was condemned to prison" and his brethren were instructed to ignore his teachings.[7] The account of Bacon's intellectual and doctrinal disagreement with his superiors reflects a later Franciscan rejection of Bacon rather than a contemporary report of events and was written in the official *Chronica XXIV Generalium* during a period of internal conflict within the Order.[8] There is no contemporary evidence to support the account of the condemnation that appears in *Chronica*. Amanda Power notes that this account gets several details wrong (Bacon was a Master of Arts, not a Master of Theology), whereas the account from Trau correctly identified subjects on which Bacon wrote and mentioned that Bacon had written his major works for Pope Clement IV.[9] Moreover, early Franciscan sources are a mixture of hagiography, legend, and *exempla* (short narratives with a moral attached to be used in sermons), and were written for multiple purposes, including to work out issues of doctrine.[10] Concerns in the fourteenth and fifteenth centuries about doctrinal orthodoxy and the intellectual activities of earlier members also appear in the Dominican Order attached to Albertus Magnus, who, like Bacon, had a posthumous reputation for making marvelous devices and for going beyond the boundaries of legitimate knowledge.[11] Indeed, it is worth noting again that Bacon wrote his major works at the request of the pope and that several prominent Franciscans engaged with and relied on his ideas during his lifetime and immediately after; the apparent injunction to avoid Bacon's teachings was either not issued or not followed. Yet this one brief remark subsequently became the source for later estimations of Bacon's character and work. As we shall see, some later writers pointed to Bacon's purported condemnation and imprisonment as evidence that he was free of papist superstition, while others suggested that these "suspected novelties" involved demonic magic or astrological determinism. Both accounts reveal the ambivalent attitude of the fourteenth century (when this account of Bacon's condemnation was written) toward novelty and innovation and demonstrate how Bacon himself became an embodiment of that ambivalence, regardless of what may have happened in 1277.[12] Bacon's association with heterodoxy and conflict with institutional authority, though likely invented, proved useful to later writers from the seventeenth century to the twentieth, as it reinforced the narrative of Roger Bacon as a misunderstood visionary in an era that had become synonymous with religious persecution and rigid orthodoxy.

Inventing and Codifying the Middle Ages

During the fourteenth century, in which Franciscan accounts of Bacon's past deeds began to appear, Italian writers began to valorize the classical past and speak in terms of "renewal" and "revival," contrasted with the preceding period of cultural darkness and decline.[13] This appeal to the past—especially the Roman past—was not new, nor was the rhetoric of renovation. The divide between the "ancient" and "new" goes back to classical antiquity. "Ancients" are classic and authoritative; exemplars that can be imitated and superseded.[14] The tribal rulers who hastened and benefited from the devolution of political power in the western half of the Roman Empire in the fifth and sixth centuries used and responded to the rhetoric of "romanitas" and renewal.[15] Later, in the ninth century, the idea that a "modern" period, looking back to ancient Rome, cast the immediately preceding period into shadow was employed by Charlemagne and the chroniclers of his reign to describe Charlemagne's restoration of Roman imperial greatness, including assuming the title of Roman Emperor.[16] Even the term "modern" is medieval. Scholars in Paris and clerks in courtly bureaucracies in the twelfth century referred to themselves as *modernus*, meaning "of the moment," and to their age as "modernity" (*modernitas*) to emphasize their differences from their predecessors.[17]

Likewise, rhetoric contrasting an age of darkness with one of enlightenment appeared long before the historical eras we now know as "the Renaissance" and "the Enlightenment." Rhetoric of ages (or states) of darkness and light became a fixture of early Christian writing about time and tense, as a way to contrast the light of Christ with the darkness of the pre-Christian world.[18] Patristic writers in late antiquity combined the teleological, vector-based structure of Christian theology (toward salvation) with older models, such as the succession of empires (Babylonian, Persian, Greek, Roman) or ages of humankind.[19] For Eusebius, historiographer, theologian, and bishop of Caesarea (d. 339), Jesus was the light that linked "before" and "after," shining backward from himself to Adam and forward from himself to the end of time.[20] For Christian writers after Eusebius, looking backward into the past became not only a way to connect with ancestral legacy, military or political glory, or moral rectitude, but also a way to envision a bright future of Christian salvation and, as with Roger Bacon, a bright future of Christian political power. Universal histories, which promised a complete history of the known world according to a Christian concept of time (beginning with Creation as laid out in Genesis) proliferated in the Latin Middle Ages in Latin and in vernacular languages.[21]

By the time that Francesco di Petracco, or Petrarch (1304–74), named his own era as one of darkness in contrast to the enlightened period of antiquity, the idea of looking backward to move forward was well established and readily invoked. As noted above, Frankish chroniclers in the ninth century viewed Charlemagne's reign and the creation of his empire as indicative of renewal and restoration; this kind of rhetoric appears whenever a ruler or dynasty seeks legitimacy and a break with their immediate predecessor. Calls for reform, renewal, and a return to a past era of learning and culture likewise sounded in the Latin Christian West in the eleventh and twelfth centuries.[22] Indeed, Bacon expressed a version of this formulation when he acknowledged the stunning technology "of antiquity," like the inventions of Alexander the Great or the telescopic lenses of Julius Caesar as proof and model for what could be made in the future. Petrarch gave the originally Christian religious imagery of light and darkness an aesthetic connotation, alongside the moral one. Like many of his earlier medieval predecessors, Petrarch found Rome to be of particular interest and importance; unlike them, he focused on Rome to the exclusion of other empires, and he wrote from the perspective of decline, rather than continuation. Other medieval historians writing from the ruins and hinterlands of what had been the western part of the Roman Empire considered the empire to have continued, though rulership had transferred (or been translated) to other peoples. This concept of *translatio imperii* goes back to the early medieval period, and was used to legitimate different rulers and dynasties that had succeeded Rome, like Charlemagne (this concept did not gain traction in the eastern half of the Roman Empire, centered in Constantinople, as imperial rule had not ended in the fifth century).[23] Petrarch, on the other hand, privileged the Republic over the Empire; named the beginning of the age of darkness as when Rome became ruled by outsiders, first by men from the provinces and then by "barbarians"; and dismissed the idea of *translatio imperii*.[24] He divided the past into two distinct periods: antiquity, the time of clarity and illumination, and his own age, one of darkness and decline.

Other writers repeated the idea that the efforts of one's forebears of a century or two ago paled in comparison to both the ancients and contemporaries. This set of ideas was taken up by Petrarch's contemporary, writer Giovanni Boccaccio (1313–75), and by others in the century after him. Lorenzo Ghiberti (1378–1455), the sculptor, writer, and painter, held that Dante, Petrarch, and Giotto had begun the rebirth of the arts out of a period of cultural decline and darkness.[25] This looking-backward-to-look-forward recalls Bacon's own Janus-faced oscillation between the devices of the ancient past (Alexander, Julius Caesar) and the brilliant future in which they once again return, even better. Italian humanists of the fifteenth century adopted Petrarch's periodization and added to it, creating the historical periodization of ancient-medieval-modern that is still going strong to this day.[26] Leon Battista Alberti (1404–72), humanist scholar and polymath, was, like Petrarch, unimpressed with the efforts of his immediate predecessors, but considered himself to be living in a new age. Giorgio Vasari (1511–74), in his influential history of art, *The Lives of the Most Excellent Painters, Sculptors, and Architects* (1550, first edition; 1568, second edition), described Giotto's painting style as a "rebirth" and marveled, "it was truly an extraordinary miracle that such an ignorant and incompetent age could have inspired Giotto to work so skillfully that drawing, of which men during those times had little or no knowledge, came fully back to life through his efforts."[27] He also characterized architecture between Alaric's sack of Rome in 410 and the late fifteenth century as "Gothic," which he defined as a barbaric degradation of classical forms.[28] For those Italian humanists who came after Petrarch and who took up his historical periodization, the rebirth or revival of antiquity was connected to the idea that a period of obscurity—a middle era—separated the two periods of light and promise.

Even as these artists, writers, and natural philosophers disparaged the ideas and aesthetics of their medieval forebears, they were influenced by them. Just as humanists had mistaken Caroline minuscule, which developed in the late eighth century, for ancient handwriting, visual artists mistook elements of earlier medieval culture as emblematic of the classical Roman past they sought to emulate: buildings and architectural features from late antiquity or the early medieval period taken as monuments from a much earlier time.[29] Ideas from medieval scholars about optics and the importance of testing and sensory knowledge influenced humanist culture. Bacon's work on *perspectiva*, which circulated as a complete text and as part of the *Great Work*, was influential in helping to spread the ideas of Ibn al-Haytham. Bacon's optics influenced vernacular courtly literature as early as the late thirteenth century (i.e., contemporary with Bacon himself); for example, his ideas are cited by French poet Jean de Meun (ca. 1240–ca. 1305) in his continuation (ca. 1265) of the earlier thirteenth-century *Roman de la Rose*. Bacon's work was also a door through which Ibn al-Haytham (as Alhacen or Alhazen) became known as an authority on optics beyond the confines of the classroom.[30] And Bacon's ideas regarding optical theory—either in his own words or synthesized by later writers, like Witelo—also find expression in the fourteenth century in other fields. Dante (1265–1321) referenced light and optics in ways that echo Bacon and Witelo throughout the *Divine Comedy*.[31] Optical motifs referencing vision, light, and reflection in Baconian terms appear in theological texts throughout the fourteenth century, including commentaries on the *Sentences* produced within the universities and the work of theologian and dissident John Wyclif (ca. 1328–84).[32] Bacon's work on color theory and perspective influenced the Florentine painter Giotto (d. 1337), whose work was later hailed by Italian humanists as the start of the rebirth of painting.[33] Indeed, the (re)development of linear perspective in painting in the fifteenth century practiced by Filippo Brunelleschi (1377–1447) and discussed by Alberti and Ghiberti drew from optical treatises by medieval authors, including Ibn al-Haytham and Bacon.[34] French theologian and natural philosopher Pierre d'Ailly (1350–1420) relied on Bacon's *Great Work* for geography and astral science; d'Ailly's *Imago Mundi* was an important source for Christopher Columbus as he planned his voyage to India.[35] The natural magic that flourished in Italian humanist culture of the fifteenth and sixteenth centuries, with a focus on *experimenta* and demonstration and an insistence that knowledge of all of nature's laws and capabilities could work wonders and bring one into a closer relationship with the divine, also owed a great deal to Bacon's ideas. As Anthony Grafton has put it, the "Renaissance magus both learned from and rejected his medieval predecessors."[36] And the engineers and artisans who produced courtly marvels, weapons, and new texts for the

Italian rulers of the fifteenth and sixteenth centuries—men like Taccola, Leonardo, Alberti, and Giovanni Fontana—"amounted to something like the realization of Bacon's project for a 'science of experience.'"[37]

In the sixteenth and seventeenth centuries the idea of a long middle period in between the ancient period and its modern incarnation became entrenched. The term "the middle age," appeared in English at the end of the sixteenth century, not long after "*medium aevum*" appeared.[38] Protestant historians in the sixteenth century, in their efforts to return to a purer version of Christianity, adopted this framework of ancient-medieval-modern (good-bad-good) as a way to dispense with the period when the corrupt Catholic Church held sway over much of Christendom. The Lutheran authors of the *Magdeburg Centuries* (1559–74), an ecclesiastical history that covered thirteen centuries, largely derided the Church from the sixth century onward.[39] In the seventeenth century Christoph Cellarius (1638–1707), a Protestant scholar based at the University of Halle, crystalized the nascent temporal framework and terminology when he published a universal history titled *Universal History Divided into an Ancient, Medieval, and New Period*.[40] Thus, "Roger Bacon, conjuring friar" is credible as an example of the decadence and depravity of some of the Church's ostensibly holiest men, yet "Roger Bacon, free-thinking visionary" is equally credible as a victim of the corrupt, dogmatic Catholic Church. In much of the seventeenth century Bacon's years as a Franciscan and his connections to Pope Clement IV were often brushed aside in favor of representing him as a victim of papist intolerance and ignorance. The preface of the English version of John Dee's (1527–1608) annotated copy of *Hidden Powers* (printed in 1659, after Dee's death) ascribes Bacon's reputation as a sorcerer to the pervasive Catholic superstition of his era, implicitly suggesting that the pope and Roger Bacon were opposed, rather than allied. "'Twas the *Popes* smoak which made the eyes of that Age so sore, as they could not discern any open hearted and clear-headed soul from an *heretical Phantasme*."[41] According to the translator, the "Age" in which Bacon lived was the middle age, a period so polluted with Catholicism and ignorance that even the wisest people could not perceive the difference between brilliant ideas and heretical illusions.

Writers in the sixteenth and seventeenth centuries declared Bacon's exceptionalism often in terms of rupture and discontinuity; these terms were largely discursive rather than reflective of reality. During this period many texts, authors, and ideas from the previous centuries remained in circulation, and were read and debated. Many of the most widely printed books in the first two centuries after the invention of the printing press were editions of medieval texts.[42] In some fields, like human anatomy, early modern reformers relied on the knowledge about the interior of the human body produced in the thirteenth, fourteenth, and fifteenth centuries, largely within Latin Christian contexts and institutions.[43]

"Roger Bacon"

Beginning in the twilight of his lifetime and for a century afterward, Bacon's influence appears largely limited to the Franciscan context, and within that context focused on optics, visual perception, and experiment (including devices). However, beginning in the fifteenth century Roger Bacon began to be replaced by different versions of "Roger Bacon": alchemist, experimentalist, and conjurer. This was due to the piecemeal circulation of parts of the *Great Work*, the wider circulation of *Hidden Powers*, and the attribution of alchemical texts to him (spuria).

Bacon discussed alchemy at length in *Hidden Powers* and in part of the *Great Work* and used it as an example of the potential of *scientia experimentalis*. A thorough knowledge of materials and their properties, gained from texts and sensory experience, combined with human know-how, could, in the case of alchemy, enable the creation of powerful curative substances that could extend human life by decades, or even centuries.[44] Yet Bacon's own ideas about alchemy were not terribly influential over the fourteenth and fifteenth centuries, except in terms of the influence on John of Rupescissa.[45] However, his reputation for extraordinary natural knowledge and his emphasis on knowledge-acquisition via *experimentum* rendered him a compelling author figure, as over twenty alchemical texts attributed to "Roger Bacon" circulated throughout the Latin Christian West from the late fourteenth and through the fifteenth centuries.[46] Ironically, Bacon himself had castigated those charlatans who did exactly this in *Hidden Powers*. "Thus, whatever they say, 'that Solomon wrote such and such,' or other wise men, must be denied, because books like this are not accepted by the authority of the church, nor by the wise, but by seducers, who deceive the world. For they themselves write new books, and multiply new inventions, as we know by experience. And then, so that they may attract men even more strongly, they affix celebrated titles to their books and impudently ascribe them to great authors."[47] One of the most important and widely copied treats rejuvenation and immortality, *The Cure of Old Age and the Preservation of Youth* (*De retardatione accidentum senectutis*).[48] Claiming Bacon as the author of numerous alchemical texts indicates that he was seen as an important author of texts on empirical subjects (alchemy, medicine) that dealt with phenomena that seemed marvelous to the uninitiated but were readily explained as the result of natural knowledge.

Bacon's work on alchemy became familiar to later authors and readers through the circulation of part of the *Great Work* and *Hidden Powers*. These were not the only instances in which he wrote on alchemy; however, those other texts did not circulate nearly as widely throughout Latin Christendom. Along with the fifth book of the *Great Work*, on *perspectiva*, the sixth book, on *scientia experimentalis*, circulated as a standalone text in manuscript in the fifteenth, sixteenth, and seventeenth centuries. However, after the invention of print, the work most frequently printed under Bacon's name was *Hidden Powers*, followed by alchemical spuria, including *The Cure of Old Age*. The first printed edition of *Hidden Powers* appeared in 1542, published in Paris, and it was often printed with the *Speculum alchemiae*, a pseudo-Baconian alchemical text. Both were printed in English in 1597 as *The Admirable Force and Efficacie of Art and Nature* and *The Mirror of Alchimy*. In contrast, the first printed edition of the *Great Work* did not appear until 1733, and it was incomplete by design, lacking the seventh and final book on moral philosophy. This omission gave the impression that the culmination of Bacon's system of the organization of knowledge was *scientia experimentalis*, a way of understanding the natural world rather than a method for knowing how to act morally as a Christian or attain *sapientia*.[49]

Due in part to the *Mirror of Alchimy* and *Hidden Powers* Bacon enjoyed a heightened reputation in the sixteenth century as an alchemist and as an experimenter, as a necromancer and a great natural philosopher. Bacon's connection with marvels, as well as his interest in astral science, including prediction, meant that he was at times presented as a suspect figure. John Bale (1495–1563), bishop, compiler, and antiquarian, noted in his first summary of major English writers (1548) that Roger Bacon was a conjurer who used evil spirits to conjure marvels. But a decade later, in the second edition of his work, he characterized Bacon as a gifted mathematician and astrologer, "but devoid of necromancy, although many have slandered him with it."[50] It is possible that Bale's concern with orally transmitted stories, here implied by the use of "slander," led to his change in view.[51] The idea that men learned in natural philosophy would be accused of practicing magic goes back at least to the eleventh century, yet by the seventeenth century it seems to have become more common.[52] In part, this is likely to do with anti-Catholic polemic that characterized the Church as hostile to scientific inquiry and the discovery of new things.[53] In addition to the examples from Dee's posthumously printed edition of *Hidden Powers* and Brown's preface to *The Cure of Old Age*, Gabriel Naudé (1600–1653) included Bacon as one of the great thinkers wrongly accused of magic in his apology. According to Naudé, it was Bacon's mastery of the science of optics and work with mirrors that accounted for his reputation as a sorcerer. Along with Bacon, Naudé defended Bacon's contemporaries, including Robert Grosseteste, Michael Scot, and Albertus Magnus.[54] As with the printer's preface to the English version of Dee's annotations of *Hidden Powers*, Naudé placed the blame for Bacon's reputation as a magician on the ignorance of the time. Bacon's insistence on the invalidity of magic in *Hidden Powers* and that magic was a sacrilegious and fraudulent activity, inferior in its effects to *scientia experimentalis*, made him a sympathetic figure to those concerned about what kind of knowledge was acceptable to a devout Christian. Bacon's impassioned condemnation of magicians in *Hidden Powers* and his careful parsing of why ancient prayers have enormous power but the incantations of unscrupulous practitioners are mere trumpery echoed the concerns of men like Bale, Naudé, and others. "His anxieties were their anxieties."[55]

The view of Bacon as a "conjuring friar," already apparent from Bale's first edition, was popularized in the second half of the sixteenth century in two important English texts. The first is the prose history *The Most Famous History of the Learned Fryer Bacon*. The earliest extant version of this text is a printed edition from the middle of the seventeenth century; however, an earlier version must have existed, given the close correspondence between the *Famous History* and the popular Elizabethan play, *Friar Bacon and Friar Bungay* (1589) by Robert Greene. Taken together, these texts mark the start of "Roger Bacon" as a comic—and quintessentially English—figure in English literature.[56] In *Famous History* Bacon's Englishness is emphasized, as he conjures marvels at the command of Edward III (1312–77) and uses his abilities to create burning glasses and optical devices to use against his enemy, the German sorcerer Vandermaster. Echoing the Franciscan chronicle by Peter of Trau, "Fryer Bacon" used his learning to create a "magick-glass," a device that shows those who peer into it what is happening elsewhere. At the end of the prose history, Bacon learns that his invention has started a blood feud between two young men. Stricken by the damage that his device has caused,

Bacon destroys the magic mirror ("the rarest in the world, made by the help of long Study and exquisite Art"), burns his books, and lives a solitary life in his cell for the next two years, before he dies penitent. Interestingly, the "Fryer Bacon" in *Famous History* lives in England in the fourteenth century—a century after Roger Bacon's lifetime—the period in which the legends of Bacon's extraordinary inventions and possible condemnations first appear. In Greene's play, "Friar Bacon" is an Oxford scholar—a "learned doctor"—who uses his knowledge to erect a brass wall that shields England from her enemies and who makes a brazen head that will reveal the future to him (he also pits his skills against the German, in this version called Vandermast) (Figure 13). However, his pursuits, though impressive and ultimately carried out in the service of England's interests, are not wholly orthodox. Bacon is a Faustian figure in the *Famous History* and in *Friar Bacon*: he consorts with a demon to build his brazen head, but he does not sell his soul to the devil in exchange for knowledge, instead gaining knowledge through long study.

The Father of English Science

During the sixteenth and seventeenth centuries, the same period in which "Roger Bacon" became a stock figure in English drama, he became credited as the "father of English science," and mobilized as a figure to bolster English claims to priority over scientific knowledge and discovery. The emphasis on Bacon as a particularly English figure is at odds with his intellectual formation in a continental context; empiricism and experimental thinking were more robust and integrated into scholastic culture in Paris in the thirteenth century than in Oxford. This widespread interest in testing and proof means that Roger Bacon was not an outlier or an exception, but rather exemplary of his intellectual and cultural formation.[57] Widespread circulation of his work on applied science and technology in the service of military aims influenced how he was presented in the *Famous History* and *Friar Bacon* as an English figure who used his knowledge in the defense of England. Other influential English natural philosophers found his work valuable and generative. Bacon's writing on *scientia experimentalis*, on optics, on devices, and his insistence on the application of natural knowledge to improve the human condition and for political gain found sympathetic adherents in the sixteenth and seventeenth centuries, especially among university-educated natural philosophers working within a courtly context.

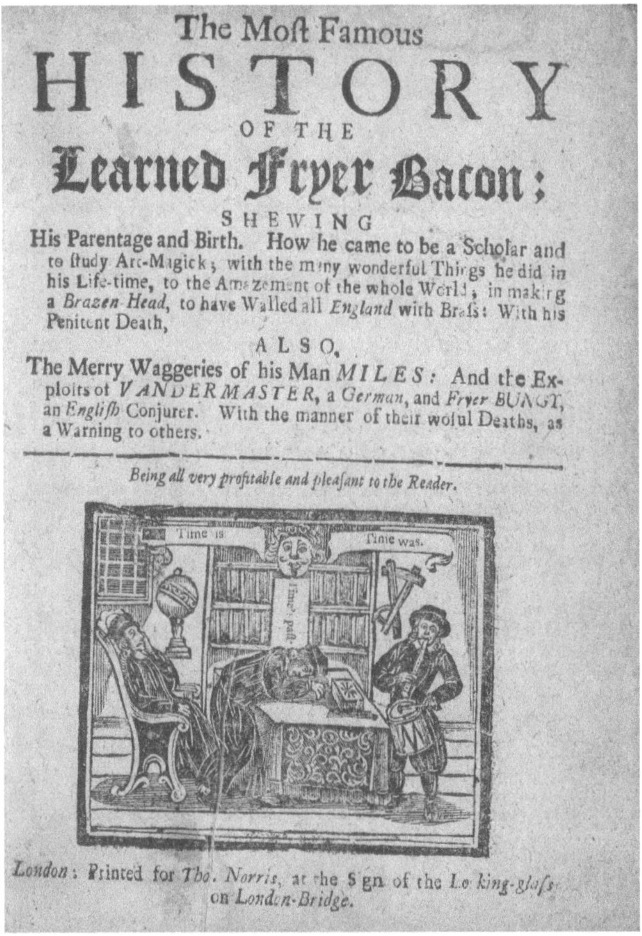

The Moſt Famous

HISTORY

OF THE

Learned Fryer Bacon:

SHEWING

His Parentage and Birth. How he came to be a Scholar and to ſtudy Art-Magick; with the many wonderful Things he did in his Life-time, to the Amazement of the whole World; in making a *Brazen Head*, to have Walled all *England* with Braſs: With his Penitent Death,

ALSO,

The Merry Waggeries of his Man *MILES*: And the Exploits of *VANDERMASTER*, a *German*, and *Fryer BUNGY*, an *Engliſh* Conjurer. With the manner of their woful Deaths, as a Warning to others.

Being all very profitable and pleaſant to the Reader.

London: Printed for *Tho. Norris*, at the Sign of the *Looking-glaſs* on *London-Bridge*.

Figure 13. The full title of this work is *The Most Famous History of the Learned Fryer Bacon; Shewing His Parentage and Birth. How he came to be a Scholar and to Study the Art-Magick; with the many wonderful Things he did in his Life-time, to the Amazement of the whole World; in making a Brazen Head, to have Walled all England with Brass: With his Penitent Death, Also, The Merry Waggeries of His Man Miles: And the Exploits of Vandermaster, a German, and Fryer Bungy, an English Conjurer. With the manner of their woful Deaths, as a Warning to Others* (London: Thomas Norris, Sign of the Looking-Glass on London Bridge, n.d.). Below the text on the title page is a woodcut illustration of Bacon, his brazen head, and his servant Miles. Bacon sleeps at his desk, surrounded by books; Miles fails to awaken him. Above Bacon the prophetic brazen head that he spent years working on speaks its lines: "Time is, Time was, Time is past." Othmer Library, Science History Institute, Philadelphia, Pennsylvania.

In the sixteenth century, as the Tudors became more interested in alchemy, they became more interested in their medieval antecedents, and in creating a particularly English alchemical genealogy.[58] Bacon is named as a great alchemist in the Ripley Scrolls of the mid-sixteenth century; his reputation as an alchemist also appears elsewhere, for example in Thomas Charnock's 1557 verse *Compendium of Natural Philosophy*.[59] In a short memoir at the end of his father's work, *An Arithmeticall Militare Treatise Named Stratioticos* (1579), Thomas Digges (1546–95) wrote that his father, Leonard Digges (ca. 1515–59), had consulted a book called *Bakons Experiments* in his own exploration of optics. Leonard Digges was an expert in applied mathematics, including surveying and ballistics, and his son linked his father's interest in optics and military application to the work of Roger Bacon.[60] There is no known extant text of Bacon's titled *Bakons Experiments*. It may be lost, or surviving under a different title, or it may be a reference to part of the *Great Work* or to *Hidden Powers*.[61] Thomas Digges also credited his father with using "perspective glasses" to see things very far away, alluding to Bacon's writing on such possibilities. This claim to the invention of the telescope served to highlight the importance of England to scientific development, a "nationalistic preoccupation that was to become central to English interest in Bacon."[62]

John Dee (1527–1608), who combined a university education with robust interest in worldly affairs and the courtly sciences, engaged Bacon's ideas about *scientia experimentalis* and the propagation of force (*species*). Dee argued for the importance of mathematics and machines to understanding nature, much like Bacon, and he saw some of the kinds of devices that Bacon had anticipated come to fruition. He was also credited with designing and building theatrical machines capable of producing many marvelous effects and was an important figure in the "Englishing" of alchemy.[63] He found in Bacon a set of ideas that profoundly influenced his own natural philosophy, and he took it upon himself to promote Bacon's work and burnish his reputation. Dee was likely introduced to *perspectiva* while studying mathematics at Louvain and later acquired several copies of Bacon's works.[64] Between 1556 and 1558 Dee focused his efforts on optics and on Bacon's works. By 1558 Dee either possessed or had consulted all of Bacon's major works, including *Great Work*, *Lesser Work*, and *Third Work*, as well as *The Multiplication of Species*, *Common Principles of Mathematics*, *Perspectiva*, and others.[65] Dee was also an intimate of Leonard Digges and raised his son Thomas in his home after Leonard died in 1559; it may have been from Dee that Leonard Digges learned of Bacon and his interest in *experimentum*, which his son Thomas later mentioned as the book *Bakons Experiments*. Dee seems to have found Bacon's writings on the propagation of light (taken as emblematic of all species), along with Bacon's ideas about composites to be productive for the development of his own ideas of the mathematical basis of celestial influence, as well as the related notion that the physical world has mathematics at its foundation.[66] He also focused with interest on those moments in his works where Bacon mentions utility (Figures 14–15).

Dee pushed back against the idea of Bacon as a sorcerer or necromancer, insisting that his ideas were important and orthodox. In 1558, the same year that John Bale published his revised opinion of Bacon, Dee wrote that Bacon "did nothing with the aid of demons, but was a great philosopher who did great things in a natural and permissible Christian manner."[67] (He had said much the same thing about himself when he was tried for treason for casting the horoscopes of the queen, Mary Tudor, and her sister Elizabeth, in 1555.)[68] Dee also annotated his copy of *Hidden Powers*, and had intended that his annotated copy be published. Dee's copy of *Hidden Powers* was printed in 1618, a decade after his death.[69] It was then later translated into English and printed in 1659, appearing as *Frier Bacon, His Discovery of the Miracles of Art, Nature, and Magick, Faithfully Translated out of Dr. Dee's Own Copy by T. M., and Never Before in English*. This text includes an apologetic preface from the translator, "T. M.," that asserts that Bacon was a legitimate philosopher and not a necromancer, and that the Catholic superstition that permeated the medieval period was to blame for Bacon's reputation as a sorcerer. Dee himself had argued that the reason Bacon had been cast as a sorcerer was due to a lack of understanding; a century later, Bacon's reputation as a sorcerer is not due to ignorance, but a result of the ideology of other Christians.

Dee was open in his admiration for Bacon's work, and he may have influenced Francis Bacon (1561–1626) to consider Bacon's work seriously. Francis, a jurist, natural philosopher, and courtier to Elizabeth I and James I, proposed and ardently championed an empiricist natural philosophy, one proceeding according to a careful method of testing hypotheses drawn from observation of natural particulars. He suggested that virtually all the ideas and texts produced by Scholasticism should be discarded except the works of Roger Bacon. He also mentioned Bacon as an example of one who used the mechanical arts to confirm theoretical precepts in *The Masculine Birth of Time* (1603).[70] In his own survey of the state of the sciences, *The Advancement*

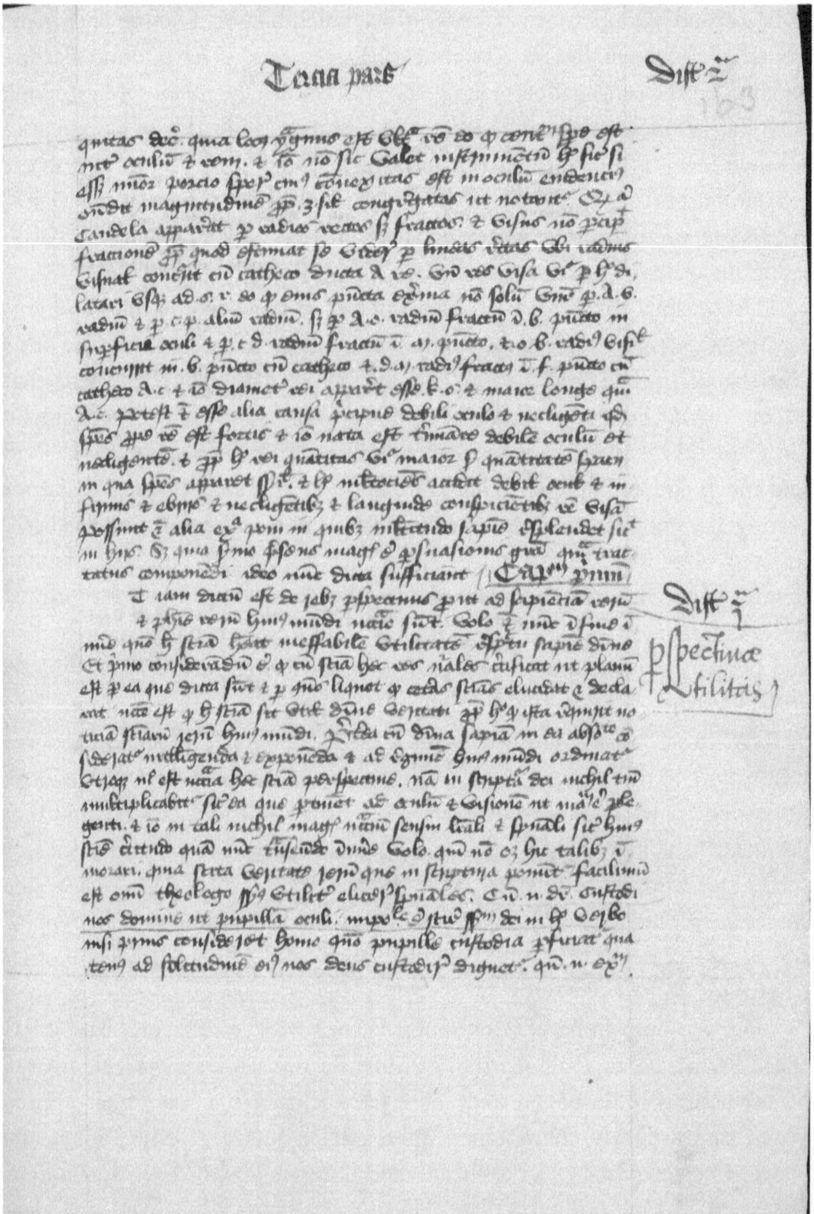

Figures 14–15. Note Dee's marginal comments on the utility of Bacon's ideas, in Bacon's *Perspectiva*, book 5 of *Opus maius*. Oxford, Bodleian Library, MS Savile 18, f. 163r, 164v.

qui sint aptitudinis et dant speciei occasionem multiplicandi se donec venerit ut fiat visio. sic corpus materiale anime sensitive ex sua proprietate et conditione administrat animam intellectivam in sua cognitione et dat ei cognitionem a parte illa qua intellectus a sensu corporali dependet. et ideo cognitio hominis quamvis sit perfecta est debilior cognitione angelica. ex hac causa et multis aliis potest comprehendi propter dictam solitudinem et languorem de homine ponitur tunc excepto beato virgine et Christo statum communem hominis et angeli. Et homo triplicem habet visionem. una perfecta que erit in statu glorie post resurrectionem. alia separata a corpore in celo usque ad resurrectionem que debilior est. tertia est in hac vita que debilissima est. et est vero per reflexionem secundum quod dicit apostolus. Videmus nunc per speculum in enigmate sed in gloria facie ad faciem. et post resurrectionem secundum plenam rectitudinem et antea in anima secundum obliquationem ab ista rectitudine: eo quod non complebitur anima plenitudine visionis antequam uniatur suo corpori. Sicut nec alias dotes habebit plenas ante illud tempus ut theologus non ignorat, quia appetitus naturalis qualis est anime ad suum corpus perfici non potest nisi in hoc. Et in statu presenti tertia est visio. scilicet vera in perfectis et fracta in imperfectis. et in malis et in religiosis dei mandata est per reflexionem secundum Iacobum. nam comparatur viro consideranti vultum nativitatis sue in speculo.

Sciendum vero quod sapientia dei operatur ad regimen mundi sit patens et utilis et pulchritudo sua ad eam videt hec istam visionem et ponam aliqua exempla tam de fractione quam de reflexione nam contra unum apparent multa et econverso infinita. Nam si videtur sint aliqui in celo sicut plures soles et lune secundum quod plinius refert de naturalibus. et hoc non accidit nisi quando vapor dispositus fuerit ad modum speculi et hoc cum sit multiplex vapor in diverso situ. et quod natura potest facere ars perficiens naturam multo magis potest idem operari. Unde possibile sit fieri et taliter poni et ordinari quod una res apparebit quot sunt volumus et ideo unus homo videbitur plures et unus exercitus plures. Et iam protracte sunt radices ad hoc. una scilicet de speculo fracto cuius perfectus recipiunt diversas divisiones [illegible] secundum diversitatem fractionum. et alia radix de aqua et speculo a quibus diversa ymago resplendet. Si ergo ordinarentur specula secundum [illegible] istorum modorum quot volumus. manifestum est quod una res apparebit in tot ymaginibus quot cupimus. Et sic pro utilitatibus reipublice et contra infideles possent huius apparitiones fieri utiles et etiam si quis novit aerem densare ut reflexio fieret ab eo possit huius utilitas [illegible]

plures soles quare appar.

utilitas

of Learning (1605) (much like Bacon's earlier *Great Work*, *Lesser Work*, and *Third Work*), Francis insisted that the mechanical arts were not only practically useful but also necessary to give an accurate rendering of natural causes. The title page engraving of this volume shows the sensible world and the world of theory clasping hands, symbolizing the establishment of Francis Bacon's epistemic method. In the *New Organon*, his system of a new natural philosophy (*Novum organum*, 1620) that was intended to be a revision of Aristotle's *Organon*, the six works on logical reasoning, Francis compared the stagnation of natural philosophy, reliant on syllogism, with the dynamism of the mechanical arts, which relies on *experimentum*. Furthermore, Francis's four "idols"—the causes of human error—resemble, at least superficially, the four causes of human error that Bacon sets out at the beginning of his *Great Work*. According to Roger Bacon, the four causes of human error are: 1) adherence to mistaken or misguided authority; 2) blind adherence to custom; 3) the prejudice of the common people; and 4) purposeful concealment of ignorance. Francis Bacon's four idols are: 1) of the tribe—errors due to human nature; 2) of the cave—errors unique to each individual due to psychology and character; 3) of the market—errors due to communication failures; and 4) of the theater—errors of dogma and custom. And Francis relied on Bacon's *Hidden Powers* (without attribution) in his late work *The History of Life and Death* (1625).[71] Francis Bacon's use of the concept of experiment (*experimentum* in Latin) reflects a narrowing of the word's earlier range of meaning, which encompassed proof and trial, but also—and importantly—included experience and knowledge gained through the senses, like *scientia experimentalis*. The inductive method of Francis relied on instruments for observation, measurement, and testing hypotheses; this is quite similar, conceptually, to Bacon's definition of the second dignity of *scientia experimentalis*: crafting instruments that can help the philosopher apprehend, via *experimentum*, natural knowledge. However, Roger Bacon's and Francis Bacon's methods do have some important differences. Roger Bacon championed *experimentum* as a way of removing doubt from conclusions arrived at through *argumentum* alone; Francis Bacon championed *experimentum* as a way of generating new theories, yet conclusions arrived at by induction are not certainties.

Although Francis is no kin to Roger, the latter has been depicted as an intellectual forebear of the former as early as the seventeenth century. *Scientia experimentalis* links them—an interest in particulars, in useful knowledge, in the secrets of nature, and in direct participation. Bacon identified instruments as part of *scientia experimentalis* and as vehicles for gaining natural knowledge (undersea exploration, optics), and he emphasized the promise of technology (or art, to use his term) to be used for practical purposes, especially in the defense of Christendom and of England. So, comparisons with the more recent, more famous Francis Bacon were perhaps to be expected. Yet Bacon was sometimes cast as the original, the exemplar, and Francis as the copy. Physician and translator Richard Browne, in his introduction to (pseudo-) Bacon's *The Cure of Old Age and the Preservation of Youth* (1684), lauded Roger, not Francis, as the inventor of a new method of scientific inquiry. "He either followed or rather invented such a Method in his Studies, as by it he discovered unknown things in Nature, and did such Wonders, that not only the Vulgar, but even some Learned Men thought him a Conjurer. Some report that he made a Brazen Head that spake and think he did it by the help of the Devil. But Albertus Magnus did the same, and Boethius the like, without any other Magick than Natural. . . . Many of Bacon's Works . . . were by some ignorant Men, that would be accounted Scholars when they could not understand them, condemned for Books of the Black Art."[72] Browne also echoed Dee and Dee's editor and translator, T. M., in refuting the idea that Bacon's work was in any way intellectually or doctrinally unsound or impious.

During the last decades of the seventeenth centuries, Bacon was remade into a figure worthy of interest by—and membership in—the Royal Society. The accounts from this period are concerned mainly with his originality and innovation. Just a few years before Browne published his translation of *De retardatione accidentum senectutis* into *The Cure for Old Age* the Royal Society expressed an interest in Bacon's work. "It was desired by that Society, that as many books, as could be procured of the said Roger Bacon should be perused; and it was wished, that they were all collected and printed, as being supposed to contain many useful things."[73] A decade later, in 1689, Robert Hooke gave a lecture at the Royal Society on the invention and use of the algorism (mathematics done with Hindu-Arabic numerals in a place value decimal system), and offered an account of the transmission of "the Pythagorian Indian or Brachman [Brahmin] Arithmeticke" from the Indians to the Arabs and then to Pope Sylvester II and after that to Roger Bacon.[74] According to his account, no one in Latin Christendom had followed Sylvester in the study of algorism in the intervening 250 years before Roger Bacon (allegedly) did so, "which was not so much to be wondered at considering the Darknesse of those times and the blacker designs of those that ruled the commonwealth of learning."[75] Hooke also repeated the story of Bacon's censure, but alongside the account of Bacon's patronage from Clement IV. "Roger Bacon underwent that grave Censure and had fared accordingly had he not been favored by [the] Pope, to whom he wrote an apology and vindication of himself."[76] In fact, Sylvester II (often called by his pre-papal name, Gerbert of Aurillac), was credited with introducing Hindu-Arabic numerals into Latin Christendom in the last decades of the tenth century, though he in fact only introduced them as counters on an abacus, rather than embedded within a decimal place system. Hooke's recruitment of Bacon into this genealogy of transmission seems to be unique, and while Bacon did use Hindu-Arabic numerals there is no evidence that he played a role in the transmission of "Indicae figurae," as they were known at the time.[77] Others at the Royal Society lauded Bacon as the inventor of the telescope, eyeglasses, and gunpowder—erroneous assertions that proved difficult to dislodge for some time.[78] John Aubrey (1626–97), in his *Brief Lives*, compiled between 1669 and 1696, wrote of Bacon, "he understood the making of optique glasses, . . . he also gives a perfect account of the making of gunpowder."[79] Claiming Bacon as a source of invention meant claiming England as the soil that had nourished his invention, and asserting English priority for Bacon's acumen.

Arabic Influence and Authority

Bacon's reputation was being made and remade in the sixteenth and seventeenth centuries against a backdrop of new tensions and interests in Arabic texts. Bacon's reference throughout his works to Arabic authors, such as Ibn al-Haytham (Alhacen), Alkindus (al-Kindī), Ibn Sīna (Avicenna), and others reflects the importance of those authors, and of Arabic textual authorities more generally, to learned culture in the thirteenth century. In some instances, as was the case with Ibn Sīna, as well as Ḥunayn ibn Isḥāq (Joannitius) and Māshā'allāh ibn Atharī (Messahallah), Arabic authors constituted the canonical works of the curriculum in Latin Christian universities, in these cases medicine and astral science, respectively. In other cases, as with Ibn al-Haytham, Arabic authors would be recognized as authorities, even if their subject matter was not formally taught in university, and their works were auxiliary to the established university curricula. The works of Arabic authors remained central to learned culture in the Latin Christian West for centuries, comprising the bulk of the curriculum in university subjects, like medicine and philosophy, through the sixteenth century, and necessitating multiple printed editions of their works in Latin translation.[80] In this period, as in the medieval period, Ibn Sina was read more widely and more carefully than Roger Bacon; for example, the latter wrote in Latin and some of his work remains untranslated, while the former wrote in Arabic and his works were translated into Latin, Greek, and Judeo-Arabic during the medieval period.

Interest in Arabic authors and their texts (in translation) remained central to the intellectual culture of the Latin West through the sixteenth and seventeenth centuries. The importance of Arabic authors and their ideas to the university curriculum, as well as in learned circles beyond the university, spurred new translations from Arabic into Latin.[81] Scholarly and religious controversies around the commentaries of Ibn Rushd (Averroes) on Aristotle in late medieval and early modern Latin Christendom are evidence enough of his importance, even without the multiple manuscript copies and printed editions of his work.[82] Over the course of the seventeenth century, interest in Arabic texts grew to include religious texts, histories of Islam, and works of literature in Arabic, Persian, and Turkish.[83] European Orientalists acquired manuscript books in these languages by the hundreds for their libraries.[84]

At the same time, many works by Arabic authors came under closer scrutiny. Some of that scrutiny sprung from concerns about the accuracy of translations of Greek texts. European humanists in the first half of the sixteenth century interested in Greek texts, especially in philosophy, mathematics, and the sciences turned to earlier Arabic translations and commentaries and began to compare them with the newly available (to them) Greek texts, unfavorably. In some cases, this had to do with concerns about corrupted translations, but in many others the objection was to the fundamental Arabic-ness of the texts and their authors. This tendency toward anti-Arabic polemic and rejection appears especially obvious in astral science, especially astrology.[85] Broadly speaking, humanism prioritized Ancient Greek over Arabic "accretions," even though the humanistic dismissal of Arabic authors had the effect of cutting off humanism from new developments in the sciences and philosophy taking place in Arabic.[86] And even the most sympathetic pro-Arabists of the humanists adopted a historical paradigm of supersession and progress from Greek to Arabic to Latin in terms of natural knowledge.[87] These humanist impulses also accelerated the Orientalization of Islam, the idea that Islam is non-Western, and that knowledge produced in Islamdom is likewise non-Western, even if produced in Iberia or Italy.[88] But there is nothing obvious or natural about this position; Iberia, southern Italy, and Sicily are all part of Europe, are all considered "Western," and were all part of Islamdom during the medieval period. Arabic authors based in places farther away from Paris or London than Palermo or Córdoba, were participating in the same Aristotelian epistemic regime as Roger Bacon. Concerns about translations, as well as humanist polemic that promoted the superiority of Greek philosophy over Arabic, resulted in an effort—successful over the longer term—to erase Arabic texts and authors.[89] The history of medieval science became focused on places in northern Europe, like Paris and Oxford, that never had direct contact with Islamicate culture. Dee's posthumous edition of *Hidden Powers*, printed in the middle of the seventeenth century, retains some references to Alhacen and Avicenna, but by that time Ibn al-Haytham's work was no longer being read, and Avicenna's importance to the medical curriculum was beginning to wane. By the time that Jebb's purposefully incomplete edition of the *Great Work* appeared in 1733, European Orientalism made it possible for European intellectuals to have fewer wrong ideas about Islam than in preceding centuries, but the Arabic authors and texts that Bacon relied on so heavily, like Ibn Sīna, Ibn al-Haytham, and al-Kindī were no longer being read by them.

The changing status of Arabic authorities and texts in the seventeenth century, along with the invention of the new historical periodization of ancient-medieval-modern and combined with the partial translations and circulations of Bacon's texts and legends about him made possible a narrative of him as a scientific genius and innovator, nurtured in England and who served England with his inventions. Avowals of novelty and innovation most often arrive within a narrative constructed with the benefit of hindsight. The "contours of newness emerge only once the anomalous is elevated, retroactively, to the status of the transformative, rather than, say, relegated to the dustbin of history as irrelevant."[90] Dee, Francis Bacon, and others translated Bacon into a figure much like themselves; an English natural philosopher and inventor of a method of natural inquiry that elevated *experimentum* alongside *argumentum* and prioritized utility, and whose works and ideas warranted interest and preservation.

PART III

Chapter 4

Periodization and Science

> There are deserts and wastelands no less of time than of place.
>
> —Francis Bacon, *Novum organum*, 1620

> The prevalence of the vague, visionary, and barren notions among these nations, cannot surprize [*sic*] us; for with regard to them we have no evidence, as with regard to Europeans we have, that they are capable, on subjects of physical speculation, of originating sound and rational general principles.
>
> —William Whewell, *History of the Inductive Sciences*, 1837

In his treatise on knowledge-acquisition, the *New Organon* (*Novum organum*, 1620), Francis Bacon asserted his own era as one of scientific innovation. This was after a long, "miserable" period, a wasteland, which had followed the only other eras of scientific discovery and progress: the Greeks and the Romans. "Nor is there any reason to mention the Arabs or the Scholastics, who in the intervening period rather destroyed the sciences with their treatises than increased their weight."[1] Francis Bacon located this chronology of scientific development in a geography—from Greece to Rome to Western Europe—that recalled the earlier tropes of the transfer of empire and the transfer of knowledge (*translatio imperii et studii*).[2] By the nineteenth century, the medieval past—though still considered fundamentally un-modern—became seen as a potentially fruitful source of national origins, as well as a source for cultural and imaginative expression in architecture, visual and decorative arts, literature, and philosophy. The terms of scholarship on Bacon—precocity, influence, and anticipation—were set in the nineteenth century alongside the institution of modern science as a set of practices and objectives, the codification of "the Middle Ages," and an efflorescence of medievalism in letters, architecture, and painting.

Bacon's exceptionalism and the period of intellectual and cultural stagnation in which he lived were well-established as facts; these facts (or "facts") became necessary to the idea of modern scientific inquiry as it was defined by historians and philosophers of science, beginning with William Whewell (1794–1866). Whewell, also a leading figure in the medieval revival of the period, presented Roger Bacon as proof that modern science—a rational, orderly knowledge-acquisition practice grounded in an experimental method—was a uniquely European Christian phenomenon, because only European Christians were capable of this kind of rational thinking and intellectual curiosity. Whewell fabricated his portrait of Bacon by suppressing the importance and sophistication of the Arabic scholars on whose work Bacon drew in the service of his Christian nationalism, and by relying on partial versions of Bacon's work.

Medievalism and the Middle Ages

Once invented, the "Middle Ages" has proved impossible to shift. A necessary corollary to "the Renaissance," the era is indispensable to claims of modernity. Bacon envisioned a future of amazing devices, but the device of periodization itself is more confining than liberating. As historian Lucien Febvre (1878–1956) put it, "We like to talk about the machines which we create and which enslave us. But machines are not only made of steel. Any intellectual category we may forge in the workshops of the mind is able to impose itself with the same force and the same tyranny—and holds even more stubbornly to its existence . . . History is a strongbox that is too well guarded, too firmly locked and belted. Once something has been put in it for safe keeping it never gets back out."[3] From initial expression to later codification and in current parlance, "medieval" has always been a term of disparagement and denigration. The imagery of the darkness and obscurity characteristic of the medieval period first articulated by Petrarch highlights both the ancient past and the "now," and links them.[4] The terminology of middle-ness indicates that the period is not now, not yet. The period lacks the vigor of the original; the excitement and potential associated with beginning; and the gravity and authority of maturity. The Middle Ages is an epoch to be endured, situated between two more interesting, influential, and important periods. It is defined by what it is not, and what it has not yet become. "Medieval" covers the conceptual distance between *now* and *then*. That distance often "functions as a kind of penal colony wherein modernity's most undesirable elements are imprisoned and conveniently redefined as 'medieval.'"[5] The adjective conveys more than a particular grouping of years in the past; the term immediately conveys the lesser value of those centuries (and the people in them) relative to "modernity."[6] Just as designating persons, places, or things as "modern" is a value judgment and an epistemic claim, "medieval" is likewise.[7] The defining aspect of "medieval" is that it is unmodern and can never achieve modernity. Whatever defines "modernity" can never be shared with anything medieval. In this way, whatever has come before modernity—the time of the "pre"—is denied the same status with whatever (or whoever) has been deemed "modern."[8] The invention of the Middle Ages in the context of the history of northwestern Eurasia made it possible for the nations that later emerged there to point to their distance and difference from their medieval past, a demarcation that Fabian echoed and linked to what he saw as the secularization of time, which had in turn been caused by the "rediscovery" of science from the Greeks in the Renaissance.[9] The explicit link with the medieval past as an immature or even inferior stage of social, economic, political, and cultural development was then used as a pretext for classifying other nations or groups as "primitive" or "barbarian." "The very moves by which European nation-based empires establish themselves across vast reaches of geographic space, constituting themselves by a simultaneous assimilation and othering of these spaces and the people who inhabit them, involves them at the same time in the invention of a complementary past *other* to themselves, a past which belongs to, but which can never be granted full citizenship in, the nation of Modernity."[10] The characterization of primitivism or backwardness was used as justification for colonization, subjugation, enslavement, and eradication. Claims of primitivism were also mobilized by European nations against other European nations; for example, the French lobbed this accusation against the Germans in the nineteenth and twentieth centuries because Germany cohered as a nation-state later than France.[11]

However, beginning in the early nineteenth century the Middle Ages acquired a positive gloss, as well. Enlightenment writers and historians, such as Edward Gibbon (1737–94), disparaged the medieval period for its religiosity and credulity. The Middle Ages were enshrined when Jules Michelet

(1798–1874) in his nineteen-volume *History of France* (1835) took up Vasari's nomenclature of "rebirth" and christened the beginning of modernity "the Renaissance."[12] Interest in the medieval period from the Romantics permeated European arts and culture and was a significant factor in political developments of the nineteenth century. Proponents of European nationalism in that period eagerly looked to the period between 300 and 1000 CE for the origins of nations, as a method to consolidate political authority or as a rallying cry for liberation. Medieval origins of a nation indicated a longstanding (perhaps perpetual) claim to territory, and they also buttressed a nation's claim to modernity. Those medieval origins—that "medieval" way of life and those "medieval" ideas—were in the nation's past. A nation's claim to medieval origins makes possible a national claim to modernity; those origins had been transcended so that the nation could become modern.

The medieval period was also a valuable resource to further colonization and imperialism, because "colonization of the past is an indispensable companion of empire."[13] This colonization and capture of both the past and other people began in the early modern period, alongside the process of defining time in terms of medieval and modern; these twin processes serve one another.[14] "The Middle Ages" acts as an imaginary place, like "India" or "Africa," in that it can be reconfigured, invented, and used to serve political ends.[15] "The Middle Ages" have also been enormously important for origin myths and for rationalizing subjugation of Native peoples and enslaved people from Africa and their descendants in the United States since Thomas Jefferson, and for finding resistance to those rationalizations.[16]

The search for origins occurred alongside the emergence of history as an academic discipline, based on the scientific approaches of philology and textual analysis. Both endeavors—nationalism and the study of history—required edited texts from the medieval period that supported the establishment of a nation-based historical narrative. This began in Germany in 1819 with the foundation, by Freiherr vom Stein, former Prussian minister of state (1804–8), of the Society for the Study of Older German History (Gesellschaft für ältere deutsche Geschichtskunde), with the purpose of publishing the Monumenta Germaniae Historica, or "Monuments of German History." Those monuments included, among others, texts from late antique African authors, papal records, Frankish statutes, and annals from the Merovingian and Carolingian dynasties (both of which were, at the same time, taken by French historians as emblematic of their own national history); in short, anything that had a link to the same linguistic group to which modern German belonged.[17] The English Rolls Series, the Patrologiae Cursus Completus (Greek and Latin writings of the patristics and medieval religious writers), and the Société des Anciens Textes Français all likewise appeared in the nineteenth century as part of the endeavor to define, categorize, edit, and organize texts from the medieval period to further nationalist aims. Scholars argued with one another about the origins of celebrated medieval epics, like *Beowulf*, *El Cid*, *Lancelot*, and *Chanson de Roland*.[18] Across Europe in the nineteenth century, interest in historical origins based in the medieval past meant that the medieval period, though still un-modern, was no longer only a barren period. The Middle Ages were still mucky, but it was the muck of primordial soup—the necessary conditions for the emergence of "France" or "English constitutional monarchy" or "the Germanic people."

"The Middle Ages" had become reified and established to such an extent that by the early nineteenth century medievalism, in the form of the neo-Gothic revival, became a dominant northern European cultural aesthetic. German Romanticism and its English counterpart both saw in the imagined medieval past an intuitive, spiritual, and ecologically focused contrast to the grim secularism of utilitarianism and the stench of industrialization. Medievalism, like neo-classicism, operates as an aesthetic pathway to ideology, and often coheres around ideas of purity, heritage, and origins. Literary medievalism and medievalism in visual arts, such as the works of the Pre-Raphaelite Brotherhood, were enormously popular throughout Western Europe (and North America) beginning in the early nineteenth century (Figure 16). Architecture often featured prominently in these authors' works, and the widespread interest in the plots of Sir Walter Scott's (1771–1832) *Ivanhoe* (1820), John Keats's (1795–1821) "The Eve of St. Agnes" (1819–20), and others kindled interest in the architecture of the medieval period.[19] The popularity of the historical works of Walter Scott, such as *The Lady of the Lake* (1810), *Waverley* (1808–13), set in the period before the union of Scotland and England, and *Ivanhoe*, set during the twelfth century, reflected and drove popular interest in medievalism in Britain and North America.[20]

Interest in narratives of historical change, in the Middle Ages, and in contemporary medievalism appear in the works of William Whewell, whose characterization of Roger Bacon and the medieval period have been enormously consequential for subsequent historiographies of science. His monumental *History of the Inductive Sciences* and *Philosophy of the Inductive Sciences* (1837 and 1840, respectively) offered both a master historical narrative for the development of modern science and a coherent account of what constitutes

Figure 16. The enthusiastic adoption of medievalism by Queen Victoria and Prince Albert ensured its entrenchment into English culture. Here Victoria and Albert appear in costume as Queen Philippa of Hainault (1313–69) and King Edward III (1312–77) at the Bal Costumé, a massive costume ball held at Buckingham Palace on May 12, 1842. This event was the first of three historically themed costume balls hosted by the royal couple. Their costumes were based on tomb effigies at Westminster Abbey, and the Throne Room was arranged and decorated to appear historically accurate. Edwin Landseer, *Prince Albert and Queen Victoria as Edward III and Philippa of Hainault*, 1842–46, oil on canvas, 143.0 x 111.6 cm. Cumberland Art Gallery, Presence Chamber, Hampton Court Palace, Royal Collection Trust, RCIN 404540. © Royal Collection Enterprises Limited 2025.

scientific knowledge, how it is acquired, and how it progresses. Whewell entered Trinity College in 1812 and soon became the center of a group of like-minded fellow students called the Trinity Circle (which included Julius Hare, Connop Thirlwall, and Hugh James Rose). Their belief in the intrinsic unity of knowledge and Christian morality brought them together, and their shared purpose was "to crush anything that challenged traditional social authority and the Church of England."[21] The conservatism of Whewell and the Trinity Circle was in part a response to the preceding decades of international upheaval and war caused by the fallout from the French Revolution and Napoleon's ambitions, and in part a response to the uncertainty in England during a period of concern about royal succession. Whewell and his circle "were on a quest to preserve the status quo and not to change the world."[22] They shared an interest in Romanticism; while an undergraduate Whewell won a gold medal for a poem in the style of Wordsworth.[23] After graduating he took up a post as a fellow at Trinity in 1817, and shortly after was elected as a fellow of the Royal Society in 1820. Whewell went on to occupy the position of Chair of Minerology before he was eventually named Professor of Moral Philosophy (1838) and then Master of Trinity College (1841). He used his prodigious intellectual gifts to further his various interests, from mathematics, to astronomy, geology, and education, as well as the history and philosophy of science.

Before he took up work on the *History* and *Philosophy* in the early 1830s, Whewell published several works of architectural history on medieval churches.[24] He traveled to the Continent in 1823 for a tour of the medieval churches of northern France, and later in 1829 to Germany; these field excursions provided the material for his book *Architectural Notes on German Churches, with Remarks on the Origin of Gothic Architecture* (1830). Later that same year he took a trip to Normandy and Picardy to examine more churches with Thomas Rickman, the English architect and author of *Attempt to Discriminate the Styles of English Architecture, from the Conquest to the Reformation* (1817). The fruits of that trip appeared in a second edition of *Architectural Notes* in 1832, which was more than double the length of the first. A third edition followed, with further material on churches in the Rhineland area, in 1839. In *Architectural Notes* Whewell defined and codified Gothic architecture as a category with internal unifying principles, distinct from Romanesque or neoclassical, and not merely a signifier of "barbarous," as earlier writers, going back to Alberti in the fifteenth century, had claimed.[25] He both established the characteristics of Gothic architecture, chiefly its principle of verticality, which is evident in the typical pointed arch, and gave an account of how this style emerged out of the devolution of the classical style and ensuing "caprice" and "incoherence" of the early Middle Ages.[26] In providing an account of how this architectural style came into being, Whewell made an explicitly historical scientific argument about causality. To realize the desire for verticality, medieval builders needed the pointed arch (and the buttress). The arch "has pillars to support its vertical, and buttresses to resist its lateral pressure: its summit may be carried upwards indefinitely by the joint thrust of its two sides. All the parts agree in this character of infinite upward extension, with an inclination or flexure to allow of their meeting at top; and thus obviously require and depend on pressures acting obliquely."[27] Whewell rejected the idea that the pointed arch was the salient feature of Gothic architecture; instead, he argued, it was the tool used to realize the idea of verticality.[28] Whewell's *Architectural Notes* were enormously influential, as indicated by the publication of multiple editions within a decade. Augustus Pugin, the architect responsible for the Gothic revival architecture of the Palace of Westminster and the tower that houses Big Ben, as well as numerous other churches throughout Britain, drew on Whewell's work on Gothic architecture for his own work, *Contrasts* (1836), in which he championed Gothic architecture and argued for a return to medieval values. French architect Eugène Viollet-le-Duc, responsible for the restorations of the cathedral of Notre-Dame in Paris, as well as the Basilica of St. Denis, Sainte-Chappelle, and Mont Saint-Michel, had read Whewell's work and used Whewell's notion of verticality in designing his own restoration of Notre Dame.[29] The work was read and discussed by members of the Camden Society, a learned society established in 1839 dedicated to the study and appreciation of Gothic architecture.

Roger Bacon and the Development of Science

In his *Architectural Notes* Whewell used the same methodology to understand historical change that he would ultimately go on to employ for other historical sciences, like geology.[30] The influential academic work of Thomas Carlyle (1795–1881), Barthold Niebuhr (1776–1831), and others testify to the appetite for and importance of historical narrative and historical causality in the early nineteenth century. The Trinity Circle were engaged and receptive readers of both Niebuhr and Carlyle; Whewell was particularly in agreement with Niebuhr on the importance of intuition and guessing in the process of discovery and was persuaded by Carlyle's theory of the "great man" or heroic individual as the driver of historic change.[31] It was also during this period that historical scientific fields, such as geology, archaeology, and philology, were established.[32] The history of science as a topic of study emerged within this broader context, to determine how scientific knowledge developed and progressed (though it was not established as an academic discipline until the twentieth century).

Although Whewell viewed Gothic architecture in a positive light, rather than simply as evidence of the continuing degradation of classical architecture, he did not ascribe the development of the pointed arch, fan vaulting, or the flying buttress to the superior intellect of medieval builders, nor did he view those developments as evidence of a robust culture of scientific inquiry and knowledge. To the contrary, the knowledge of medieval builders may have enabled impressive engineering feats, but, according to him, "such knowledge is manifestly not of the nature of that which we call *science*."[33] Just a few years after his *Architectural Notes*, Whewell began work on the *History of the Inductive Sciences*, which was first published in 1837 and appeared in multiple printings and editions for the next several decades. The *History* was the foundation for his account of the development of scientific knowledge, *Philosophy of the Inductive Sciences* (1840); to understand and explain his theory of induction Whewell first explored the history of science, from antiquity to the recent past. Exploring the development of inductive reasoning in the past was a way for Whewell to test his theories of how scientific knowledge is made and how it had progressed (or not) since antiquity.[34]

Whewell aimed to establish a unity of knowledge, with natural philosophy and Christian morality in a mutually reinforcing relationship. Like both Francis and Roger Bacon, Whewell's theory of scientific knowledge-production involved using empiricism in tandem with theoretical or abstract knowledge, thus resolving the "fundamental antithesis" between these two epistemic methods.[35] Like Francis Bacon, Whewell held that the method of inductive reasoning was more than simply enumerating facts or particulars, but, rather, putting together different facts or instances and making educated guesses as to what connects them. Whewell called this act of superimposing a unifying idea, one that can be expressed as a universal law or theory, "colligation," and it is the first step in scientific discovery. After the concept uniting the instances is expressed as a universal, the universal is then extended to an entire group. According to Whewell's theory of colligation, discoveries do not require the production of new facts. Discoveries are made when facts are connected by the idea that correctly unifies them.[36] The ideas themselves (like number, duration, or size), Whewell believed, are fundamental to the human mind, and are echoes of the categories and ideas that God used during creation.

Presenting a coherent explanation of how scientific discoveries happen and how scientific knowledge had developed into a scientific method enabled Whewell to assert—and to demonstrate—how Christian moral philosophy and natural knowledge are inextricable.[37] For Whewell, progress in knowledge rested on a foundation of Christian morals and he, like many, believed that England had been chosen to spread Christianity and civilization.[38] Whewell also believed that England was uniquely important to the development of science, because of the advances made by Francis Bacon, Robert Boyle, Isaac Newton, William Herschel, and even Roger Bacon.[39] Whereas scholars of the sixteenth and seventeenth centuries, such as John Bale, Gabriel Naudé, Richard Browne, and John Aubrey made Bacon into a standard-bearer for English science, despite his Continental ties, for Whewell Roger Bacon's "Englishness" could be transmuted into Britishness in the service of the British Empire. In Whewell's reasoning, Bacon's priority in the development of science demonstrated the inherent modernity of Britain and its Christian morality, both of which made Britain's colonization of other nations and peoples inevitable and providential. The similarities between Whewell's theory of knowledge production and those of Roger and Francis Bacon also provided a kind of proof through repetition that scientific thinking was inherently English and Christian. Both Whewell and Bacon wrote during periods of intense political conflict and social unrest, both believed in the unity of knowledge in a divinely created universe, both saw the necessity of balancing *argumentum* with *experimentum*, and both believed in the inextricability of Christian morals and natural knowledge. Yet where Bacon sought to overhaul the existing university curriculum, Whewell sought to maintain it.[40]

The Middle Ages and Roger Bacon figure as enormously important in Whewell's account of the history of induction, despite Whewell's characterization of the intellectual life and culture of the medieval period as a wasteland. The epigraph to Book IV, on medieval science, is taken from Alexander Pope's satirical epic poem, *The Dunciad* (1728–43) about Dulness [*sic*] overcoming Truth, Philosophy, Physics, Mathematics, and Metaphysics.[41] Technology (or "art," as Whewell called it) emerges before science. The apprehension of principles (or natural laws) through practice necessarily precedes theoretical discovery. The many inventions of the Middle Ages (in addition to Gothic architecture he listed parchment, paper, clocks, better processes for making glass and steel, algebra, and the mariner's compass) may be impressive, especially in such an intellectually stale period, but they "are no evidence that the sciences then existed; but only that those powers of practical observation and practical skill were at work, which prepare the way for theoretical views and scientific discoveries."[42] These medieval inventions did not elucidate any new theoretical knowledge, or any previously unknown natural principles. "What mechanical principle unknown to Archimedes was displayed in the printing-press?"[43] Know-how is not the same as theoretical understanding. If ancient Greece was "the period of the first waking of science," the Middle Ages was "that of its mid-day slumber."[44] To Whewell, who saw his *History* as "fairly exhibiting the progress of scientific truth," the Middle Ages were "the stationary period," a "long and barren" interval "between the scientific activity of ancient Greece, and that of modern Europe."[45] This interval was interrupted only intermittently by the technological inventions of the period, those "small successes of Art."[46] The cause of this supposed intellectual torpor was the "servility which had yielded itself to the yoke" of religious dogma, coupled with an "intolerant disposition" and an "enthusiastic temper," which led scholars and philosophers to focus on the works of previous authors to the exclusion of observation.[47] Therefore, according to Whewell, "experimenters were replaced by commentators; criticism took the place of induction; and instead of great discoverers we had learned men."[48]

Like Francis Bacon before him, Whewell viewed the sole exception to this period of intellectual servility and mere enthusiasm to be none other than Roger Bacon. According to Whewell, only Bacon, alone of his contemporaries or those who came before and after him for centuries, apprehended and argued for the importance of experiment and skepticism for authority. "Roger Bacon's works are not only so far beyond his age in the knowledge which they contain, but so different from the temper of the times, in his assertion of the supremacy of experiment, and in his contemplation of the future progress of knowledge, that it is difficult to conceive how such a character could then exist."[49] For material on Bacon, Whewell had the earlier estimations of his work to go by, as well as printed editions of *Hidden Powers* and Jebb's 1733 incomplete edition of the *Great Work*.[50] The latter was especially important to Whewell, as the printed edition made engagement with

Bacon's ideas in the *Great Work* significantly easier to undertake. Whewell likened the *Great Work* to a combination of Francis Bacon's *Novum Organon* (1620) and Diderot's *Encyclopédie* (1751–1766), both major works of epistemic reform. Whewell described Bacon in terms of anachronism and precocity ("far beyond"), the terms in which he and his scholarship have been defined ever since. Whewell also praised Bacon for his efforts at educational reform, his interest in advancing scientific knowledge, and his focus on both mathematics and experiment.[51]

To Whewell, Bacon's definition of *scientia experimentalis*, with its emphasis on the necessity of both *argumentum* and *experimentum* to acquire natural knowledge, was very similar to Whewell's own idea that scientific knowledge requires both universal idea, or theory, and particulars. Whewell used this similarity between Bacon's ideas and his own to buttress his claim of Bacon's precocity and visionary tendencies. Bacon's insistence on the importance of these two methods to knowledge as a route to moral philosophy and *sapientia* may have also resonated with Whewell, who saw Christian morality as a necessary component of natural knowledge. Natural knowledge through induction was "a good thing," according to Whewell, but it needed to be reconciled with Christian truth.[52] In *Philosophy of the Inductive Sciences* Whewell asserted the intimate and necessary connection between knowledge from the sensorium (facts) and abstract ideas or theories. Science progressed only when both these elements were present in the appropriate measure. Overreliance on empiricism had prevented botany from developing as a science, whereas enthusiasm for theoretical knowledge and received wisdom characterized "the stationary period" of the Middle Ages.[53] Bacon, on the other hand, "directed his improvement to the two essential parts of all knowledge, Ideas and Facts, and thus took the course that the most enlightened philosophy would have suggested."[54] Whewell framed what he considered Bacon's similarity to himself as evidence of Bacon's enlightenment in an age of darkness.

However, the intellectual kinship that Whewell claimed with Bacon meant only that Bacon was just as "enlightened" as Whewell, not that Whewell was as "medieval" as Bacon. Whewell clearly evaluated the past, Roger Bacon included, according to the values and interests of the present, including his own definition of science, the role of experiment, and the kind of character, intellect, and embodiment that he believed was required to produce scientific theories. Furthermore, Whewell's (and his contemporaries') ability to appreciate the importance of Bacon's ideas was then taken as another indicator of the scientific progress between the thirteenth century and the nineteenth. Modern minds could understand Bacon's ideas, but Whewell and others give no sense that they thought Bacon could understand theirs, despite the similarities.

The intellectual torpor of the Latin Middle Ages was, according to Whewell, a result of a culture in the thrall of religious dogmatism and mysticism and hostile to rationality and curiosity. Whewell, a committed Anglican Christian (and Master of Trinity College, a member of the clergy, as was required by the institution), held that religious belief and scientific thought were compatible, and that the latter required the former to be true. But during the Latin Middle Ages Christian people's minds were clouded by papist superstition, intellectual rigidity, and spiritualism. Others, either non-Christians or people living beyond northwestern Europe, were simply incapable of producing scientific knowledge because they lacked curiosity, intellectual sophistication, or the ability to throw off the yoke of religious belief. Whewell was especially dismissive of Arabic authors (and there is no mention of work by scholars in the Byzantine Empire), whom he characterized as non-European (regardless of where they lived or were from), notable only for their "intellectual servitude."[55] According to Whewell, scholars and philosophers working in Islamdom were unable to surpass Greek science because they lacked "individual consciousness, independent volition."[56] In addition to Muslims, according to Whewell (he conflated people living under Muslim political authority with Muslims and Arabs), "the Jews, the Indians, the Siamese, and the Chinese" also lack any ability to do science. They could not sustain systematic gathering of facts, nor could they generate theoretical principles. "The prevalence of the vague, visionary, and barren notions among these nations, cannot surprize [*sic*] us; for with regard to them we have no evidence, as with regard to Europeans we have, that they are capable, on subjects of physical speculation, of originating sound and rational general principles."[57] A few "Arabian" astronomers had been able to make incremental improvements on Greek astronomy, but they failed to lead to any new theories. And even though they did a few experiments, they still failed to understand "several of the sciences which the Greeks had founded," namely mechanics, hydrostatics, and harmonics.[58] In the case of Arabic texts and their authors, the "Arabians" were "uncivilized" and, therefore, had "not been duly prepared rightly to enjoy and use the treasures [of Greek science] of which they became possessed."[59] Their slavish obedience to military command and reverence for the Qur'an meant that they "merely translated" and "swelled the vast and unprofitable herd" of Aristotelian commentators.[60] Arabic authors had more access to previously established scientific knowledge than their contemporaries in the Latin Christian West, but according to Whewell they failed to engage with it or to develop their own theories of experimentation. He took this as further evidence that Arabic authors were not intellectually as capable or inventive as Roger Bacon, or the natural philosophers of the sixteenth and seventeenth centuries, rather than as evidence of his own ignorance, or the ignorance of those who came before him.

Instead, Arabic writers merely preserved Greek scientific ideas until Christian Europeans like Bacon could turn them into something that looked like modern science. That Bacon did so is presented as evidence of the uniquely European capability for producing scientific knowledge.

> That [Bacon] received much of his knowledge from Arabic writers, there can be no doubt; for they were in his time the repositories of all traditionary knowledge. But that he derived from them his disposition to shake off the authority of Aristotle, to maintain the importance of experiment, and to look upon knowledge as in its infancy, I cannot believe, because I have not myself hit upon, nor seen quoted by others, any passages in which Arabian writers express such a disposition. On the other hand, we do find in European writers, in the authors of Greece and Rome, the solid sense, the bold and hopeful spirit, which suggest such tendencies.[61]

Whewell's reading of Bacon's attitude toward Aristotle reveals more about Whewell and his ideas about what constituted independence of mind than anything about Roger Bacon. Bacon's interest in elucidating or clarifying Aristotle stemmed from his concerns about poor translations that had corrupted the original meaning or from the need to update Aristotle on subjects unknown to him. Bacon echoed others, like Grosseteste and Frederick II, on the importance of emending Aristotle; slavish dedication to Aristotle is not found in thirteenth-century Latin learned culture. Moreover, Whewell was either ignorant of Bacon's interest in the *Secret of Secrets* (which was believed by Bacon, and many others, to be a genuinely Aristotelian text) or ignored it. Either way, Whewell overlooked Bacon's intense engagement with and learning from the *Secret of Secrets* in favor of depicting him as opposed to Aristotelian knowledge, as though Bacon had lived in the seventeenth century, rather than the thirteenth—the same kind of temporal dislocation and equivocation that situated Bacon during the fourteenth-century reign of Edward III in the *Famous History*.

According to Whewell, Bacon's work superseded—in importance, complexity, and modernity—that of Arabic authors who influenced him. Earlier efforts in European intellectual culture to downplay or entirely omit the work of Arabic scholars rested in part on ideas about the superiority of Greek philosophy, and concerns about earlier translations. Overt Islamophobia and anti-Arab sentiment, though present, were not hallmarks of Latin humanist culture writ large.[62] By Whewell's time, however, authors like Ibn Sina and Ibn al-Haytham were no longer being read by scholars in Europe.

However, Whewell had no desire to find experimental thinking in the works of Arab scholars, as to have done so would have disproved Whewell's view that Muslim people, along with Jewish people, and people from India and China, among other places, inherently lacked the inclination and intellectual ability to produce science, despite all the inventions they could claim.[63] "The Arts may have had their birth in all parts of the globe; but it is only in Europe, at particular favoured periods of its history, which has ever produced the sciences."[64] Downplaying, ignoring, or demeaning the efforts of Arabic scholars, insisting on conflating Arabic authors with Muslim authors, insisting on their otherness, casting them as "mere translators" and repositories of Greek philosophy and science, bolstered Whewell's argument for Bacon's exceptionalism and, by extension, European (especially English) exceptionalism. This was his goal. Whewell wrote to a correspondent in 1837, shortly after the *History* had been published, that he would have little to correct in terms of the history of optics for future editions of the *History* (and for his *Philosophy*), "except, perhaps, the question between Ptolemy the Greek and Alhazen the Arabian, and I shall be glad if I find the Arabian discovered nothing, for I have a strong *European nationality*."[65] He repeated this sentiment in a published letter to the editor of the *Edinburgh Review* in January 1838 on the subject of Alhazen's (Ibn al-Haytham's) contribution to optics, arguing that further research into "the Arabian" would likely confirm his view that Alhazen "derived all his experiments from the Greek." Furthermore, "this will confirm an opinion which I have strongly insisted upon—that the Arabians made very few original advances in science. This I shall be pleased to find . . . I cannot divest myself of a feeling of *European nationality*."[66]

Whewell's desire to present a narrative of scientific development that privileged Europe drove his decisions about what to include and what to leave out of his account. He had pages of notes on Alhazen's work on optics and his interest in experiment from his research for *History*. These notes, which date from 1831 and 1832, contain reference to Alhazen and his use of inductive reasoning for his theory of intromission, his influence on Bacon and Witelo, and praise for his ideas. "Alhazen appears to be a good thinker. His reasoning is mostly right."[67] In another set of notes, from the same period, on what he called "the beginning of experimental philosophy" (*initia philosophia experimemtalis*), Whewell noted Alhazen's use of instruments in his work on optics. "Alhazen describes an instrument to measure reflexion and refraction and explains what the 'experimentator' is to do with it."[68] Elsewhere in his notes on the history of medieval astronomy from the late seventh century to the early fifteenth century, Whewell only listed Muslim scholars and events significant to Islamic history: the siege of Constantinople, the caliphates of "Al-Mansure" and Harun al-Rashid, and the dates of Thabit ibn Qurra, Alfraganus (al-Farghani), Charles Martel, Alhazen, and Averroes.[69] He clearly understood the importance of the work produced by scholars working in Arabic, many of them Muslim, in the fields of astronomy, optics, and even experimental science. Whewell left this material out of the *History of the Inductive Sciences* because it challenged his ideology of Christian nationalism and his narrative of Bacon's invention and the English origin of modern science.

Over the succeeding decades Whewell continued to extol Bacon's exceptionalism and his innovative thinking. In the third edition of *History*, he included extensive material on Roger Bacon reprinted from the *Philosophy of the Inductive Sciences*, because of the role that Bacon played, as an Englishman, in the history of science. Not everyone was willing to grant to Bacon the inventions that Whewell assigned him. In 1859 Augustus de Morgan (1806–71), the British mathematician and former student of Whewell's, wrote to John Herschel, "I do not believe Roger Bacon had a telescope. He had spectacles and speculated on compound spectacles. He says a boy would appear a giant—proof enough he never had seen through a telescope."[70] Nevertheless, the overall picture of Bacon and his role in the development of science presented against a backdrop of intellectual stagnation, drawn by Whewell, had taken on its unmistakable contours.

Typological Science

For Whewell, like others of his period, the Middle Ages are both barren desert and primordial soup. The fact that Bacon, a far-sighted visionary, existed within this stagnant period was proof of his extraordinary gifts, but more broadly demonstrated the supposed uniquely European aptitude for science. It was during the medieval period that "Europe descended, in intellectual character, to the level at which the other parts of the world have always stood."[71] Embedded within a larger culture that prioritized submission to authority and adherence to dogma, Bacon's interest in *experimentum*, his supposed dismissal of Aristotle, and his purported clash with Church authorities (Whewell asserted that Bacon had been incarcerated for his views), were all the more remarkable, and were taken as evidence of Bacon's uncharacteristically modern disposition in an un-modern time.[72] If he could emerge out of the intellectual stasis of the period, supposedly completely out of step with his own age, it proved that only Christian Europeans—especially Englishmen—had the capability to do science, because, according to Whewell, the medieval period was the only time at which the civilization of Europe "descended" to the level of everywhere else. Both "the Middle Ages" and Bacon's ideas are central to Whewell's argument about the development of modern science being specifically and only a Christian European phenomenon, and, moreover, that the development of modern science could only ever have been a Christian European phenomenon. Finally, Bacon's clear interest in using scientific knowledge for political ends, either on behalf of strengthening Christendom (in his own writings) or England (in his own writings and the later legends surrounding him), or conquest, or both, matched the interest of Whewell and his contemporaries—members of Britain's ruling elite during the height of British Empire.

The framework for understanding and portraying Bacon that Whewell and those who came after him employed is typological, and it is associated with medieval Christianity. Suggesting that Bacon was, somehow, "ahead of his time" in his ideas about the acquisition of natural knowledge, and that his abilities and ideas demonstrate the Christian European character for and origin of modern science requires a relationship between the past and the present in which events in the past prefigure later events. The "type" can be considered the ideal, as with Platonism and Neoplatonism, in which the ideal forms are the type. However, beginning in late antiquity, typology became particularly associated with Christian theology, and was a way of reading the Hebrew Bible and the Gospels. In the first centuries after Jesus, theologians grappled with the relationship between Judaism and Christianity, and between the Jewish scriptures and the new testaments that arose in the first centuries after Jesus's execution. Unlike Judaism, Christianity is a universal religion. Christian theology maintains that it has always existed since creation, latent, because Christianity was always going to appear "in the fullness of time"; Judaism has a specific origin story long after the moment of creation. But early Christian theologians needed to reconcile Christian universalism with what had come before Jesus. How could Christianity exist before Christ? Following Paul, early Christian writers wrestled with this question and came to different answers, including outright rejection of the Jewish scriptures, but gradually cohered around typology and replacement. Christianity had superseded Judaism; Judaism was a lesser, incomplete, or immature form of Christianity.[73] Eusebius of Caesarea (ca. 260–339), the historian, polemicist, and eventual bishop of Caesarea, reconciled the time before Jesus with the time after by viewing the "before" as the immature or incomplete version replaced by the fully realized "after."[74] Events in the Jewish scripture (now known as the Old Testament), like the sacrifice of Isaac (the *akedah*), are then interpreted as prefiguring the fulfillment of those events in the Gospels, like the crucifixion of Jesus. Likewise, the three days that Jonah spent inside the belly of the whale are a reference to Jesus spending three days in his tomb after the Crucifixion. According to this method of interpretation, the most important events in the Gospels correspond to prior events recounted in the Hebrew Bible that occurred before the birth of Jesus; however, those prior events are the incomplete harbingers of what is later fully realized once the conditions for possibility have been met.

This type of exegetical reading was widely practiced throughout the Latin Middle Ages, especially from the twelfth century onward. For those who were educated in schools, and, later, universities, the "Regular Commentaries" (called the *Glossa ordinaria*) on the Christian Bible provided a full introduction to this exegetical method by using examples from the writings of the Church Fathers in which the typology between Old and New Testaments is clearly spelled out.[75] In the elite setting of the court, especially in France, this kind of figural reading (as it is also called) was carried out within the elaborate and richly illustrated moralized Bible manuscripts of the thirteenth century (Figure 17). These bibles made the conceptual connections between Abraham and Jesus (in founding the Church) or Moses and Jesus (in leading believers to salvation) clear in a carefully laid out arrangement of images. Paired sets of images conveyed correspondences between the Old and New Testaments that reinforced the typological relationship between Judaism and Christianity.[76] Christian time is teleological: from creation and then the absence of Christianity, to the Incarnation and Resurrection, and then to the Second Coming. The narrative arc of Christianity mimics Aristotle's metaphor for human development—the acorn that becomes the oak tree.

Both are stories about linear potentiation, about development from an immature early period to maturity and the full realization of original potential. This typological reading is teleological and progressive and views the past only in terms of the demands of the present and an imagined future.

Both the Christian doctrine of supersession and Whewell's argument for the emergence of modern science divide time into epochs of "before" and "after," and leave some groups immured in the age of "before."[77] The logic of replacement keeps Jewish people in the past as long as they fail to be Christian. By remaining Jewish even after the existence of Jesus, according to replacement theology, Jewish people remain in an earlier, immature part of human history and are unable to move forward into the present and future. Whewell held that Jews are one of the groups that did not and could not produce scientific knowledge, alongside "Arabians" (by which Whewell meant Muslims, ethnic Arabs, and non-Jews living in Islamdom), and all other non-European, non-Christian people. Moreover, his depiction of intellectual and scientific culture in medieval Islamdom as one of merely translation and conservatorship defines that entire enterprise only in terms of its relationship to Greek texts: a "golden age" from the Greco-Arabic translation movement in the late eighth and ninth centuries to the Arabo-Latin translation movement in the twelfth and thirteenth centuries. Once the "treasures" of Greek philosophy, medicine, and mathematics had been conveyed into Latin so that Latin Christian scholars could properly engage with and understand these texts, as this narrative goes, science in Islamdom began its inexorable decline.[78] Although Whewell was a sympathetic reader of Niebuhr and the cyclical vision of history put forth by the Liberal Anglicans, he had much in common with the Whig position, including attunement to the past, and the notion of progress.[79]

Historians of science beginning with Whewell applied this deeply medieval, Christian hermeneutic to support a narrative of historical and scientific progress, one that privileged "Christian Europe" or "Western civilization" as the only inventors (and the only possible inventors) of science. Despite a clearly apprehended and articulated similarity between Bacon and later thinkers, later thinkers, like Whewell, do not claim to be "like Bacon" (medieval). Rather they say, "he is like us" (ahead of his time) because to be medieval—*except in the case of Roger Bacon*—is to lack the character and intellectual capability to do science. Bacon's exceptionalism affirms the "Western" origins of modern science, and Bacon is the exception that proves the rule; in this case, that the Latin Middle Ages were a period of intellectual stagnation.

Figure 17. A page from a *Bible moralisée* showing the correspondences between the Hebrew Scriptures and the Gospels. Made in the Île de France, second quarter of the thirteenth century. London, British Library, MS Harley 1527, f. 4v.

The early nineteenth century saw the entrenchment of the historical periodization that first emerged several centuries earlier. The "Middle Ages" as an invented historical era became a useful tool in the nationalist and imperialist projects of European nations and the United States in the nineteenth century, as *fons et origo* of the nation and as the un- or underdeveloped past that had been outgrown in service to modernity. At the same time, interest in the period—in medievalism—operated as a rebuke to modernity and a route back to a more innocent time. Whewell, at the nexus of the history of science and the aesthetics of medievalism, recruited both the invented figure of "Roger Bacon" and the invented historical period of "the Middle Ages" in his history of scientific progress, and in the invention of modern science as a method of knowledge-production rather than a type of knowledge. Whewell looked to Bacon as the origin and proof of the Englishness of modern science and took as determinative of Bacon's genius and modernity his similarities to Whewell's own views. *He is like us* here functions to colonize the past with a prophetic figure who pointed to a more realized future. This narrative structure of Whewell's historical chronology is borrowed from Christian theology. The centrality of typology to Whewell's argument aligns with his views on the special role of the Christian church, especially in England, and on the unity of natural knowledge and Christian morality. Christian theology from late antiquity and the medieval period, and the idea of "the Middle Ages" itself, are constitutive of modern science.

To tell this story of the figure of "Roger Bacon, the first scientist" and the development of science in Christian Europe Whewell not only relied on Christian ideas about time and history, but he also downplayed or omitted the influence and importance of Arabic texts written by scholars in Islamdom. This was, as he put it, in the service of his "European nationalism" and his belief in the role of Britain to spread Christianity and modernity. Bacon's exceptionalism (for his time) and his exemplarity (of modern scientific rationality) were constructed in the service of a narrative of European (especially British) exceptionalism and exemplarity. Whewell's views about who has the capacity to think experimentally and carry out science and his narrative about where science first emerged both participated in and reinforced contemporary views about scientific racism, specifically the idea that white people from northern Europe were more intelligent and more human than people from other places.

In addition to codifying the structure of scientific process according to Christian typology, Whewell also affirmed the view that the history of science is the history of scientific discovery and the individuals who make those discoveries. Yet Whewell made no discoveries of his own and his name is not widely known; Francis Galton, writing in 1869, held up Whewell as the example of a brilliant mind of his age, destined (or doomed) to be forgotten.[80] Although Whewell may no longer have been mentioned by name at the turn of the twentieth century, the historical narrative he created was widely taken up, especially in the Anglophone world. As we shall see in the next chapter, the narrative of scientific progress and periodization proved so durable that it could accommodate the removal of overt Christian morals from the story, despite the centrality of Christian morality to both Bacon's and Whewell's views on natural knowledge and scientific practice.

Chapter 5

The Structure of Scientific Progress

> [In] all fairness, when we speak of the Baconian reform of science, we should refer to the forgotten monk of the thirteenth century rather than to the brilliant and famous chancellor of the seventeenth.
>
> —Robert Adamson, *Roger Bacon*, 1876

Over the second half of the nineteenth century Whewell's narrative of scientific progress and the shift to modernity was further entrenched and reinforced by historians such as Jules Michelet and Jacob Burckhardt, who codified the terms of the Renaissance as a break with the Middle Ages and the beginning of modernity.[1] And Whewell's characterization of both the Middle Ages and Roger Bacon were taken up enthusiastically by other historians of science in the nineteenth century, even those who offered a different view of Bacon's relationship to religion. Over the course of the late nineteenth and twentieth century, the characterizations of Bacon, the time in which he lived, and his relationship to modern experimental science, all codified by Whewell, remained largely intact. Science became secularized and so did Bacon, but the same typological reading of potentiation and progress, which had been inherited from medieval Christianity, remained at the heart of the history of science.

The First Scientist

The successes of Whewell's *History* and *Philosophy* and the impact of his depiction of Bacon drove later interest in Bacon's contributions to science. The link between Bacon and inductive reasoning likely recommended Bacon as a subject to two scholars and professors of logic in the second half of the nineteenth century. Émile Charles, professor of logic at the Lycée de Bordeaux, writing in 1861, considered Bacon to be an exemplary figure of modernity, though trapped outside of it. Alive in a context in which freedom of inquiry was impossible, and religious ideology forced science to become cautious and furtive, only Bacon "claimed the right to think for himself," and championed a method "without which all progress becomes impossible."[2] Even though Charles claimed to have based his study on Bacon's manuscripts, he de-emphasized Bacon's close ties to the pope and his focus on Christian moral philosophy in favor of a progressive view of history and the character of "Roger Bacon, free-thinking individualist."

In attributing the repression of science to religious ideology Charles not only erased Bacon's own adherence to religious ideology but also made a claim based on the mistaken (though still widespread) idea that Christianity, or Christian ideology, during the Latin Middle Ages, was a uniform, undifferentiated entity. It was not. Throughout Latin Christendom—even in just the thirteenth century—there were multiple ideas about what "the Christian Church" was or should be, about what constituted living a life that honored Jesus and his church, and about the orthodox celebration of different Christian rituals and rites. But throughout the Middle Ages[3] there was so little coherence that at times multiple popes occupied the position as anointed Bishop of Rome, dog-headed giants and actual dogs were named as saints, priests had wives and concubines, monks tried to kill their abbots, kings and emperors fomented rebellion against bishops and popes and vice versa, some popes were believed to be diabolical necromancers, and groups of laypeople claiming to follow Christ's teachings were persecuted and murdered by people acting in the name of the Catholic Church.[4] Different groups disagreed about major points like the date of Easter; parish priests varied widely in their literacy and knowledge of Latin; Christian leaders openly defied public sentiment and political pressure (for example, as when bishops in the Rhineland shielded the Jews in their towns from murderous pogroms in the 1190s).[5] These are only a few examples to give a sense of the complex, nuanced realities of the medieval Latin Church and to demonstrate that to speak of "religious ideology" as a coherent entity in this period says less about the period and ideas under discussion than it does about the assumptions underlying that statement in the first place.

Charles's intellectual biography of Bacon did not pay close attention to his method of knowledge acquisition, at least according to Robert Adamson, instructor of logic at Manchester. To fill out the picture of Bacon, Adamson prepared and published a lecture in 1876 partly in response to this perceived lack in Charles's study. "M. Charles, however, does not seem to me to have appreciated fully Bacon's attempted reform of scientific method, the most valuable part of his philosophy."[6] Adamson lauded Bacon as a revolutionary thinker, pursuing freedom of thought, whose works had been forgotten during his lifetime and misunderstood more recently.[7] He denied Bacon's religious vocation, saying of Bacon's identity as a friar, "Why he took this step it would be hard to say. We can see why he should have preferred the Franciscan to the rival order, the Dominicans; for the Franciscans generally were distinguished for the attention bestowed by them on practical science. But that he should have subjected himself to the rigid discipline and narrow intolerance of any order is surprising."[8] In Adamson's account, Bacon's clash with Church authorities is evidence of his commitment to free inquiry, and to his own theory of knowledge. Adamson followed Jebb's printed edition of the *Great Work* and omitted the seventh and final book of that work—the section on moral philosophy—and represented the sixth book on *scientia experimentalis* as the pinnacle of Bacon's theory of knowledge, thereby downplaying Bacon's commitment to Christianity.[9] Adamson repeated this in his entry for the *Dictionary of National Biography* (1900). This mischaracterization of the organizing principle of the *Great Work* fit Adamson's narrative that Bacon anticipated a secular experimental method.

One significant change between Adamson's 1876 lecture and publication and his later *Dictionary of National Biography* entry illustrates clearly how Whewell's narrative had taken hold, even when secular rationality had replaced Christian morality. In the earlier account, Adamson devoted several pages to Bacon's debts to Arabic Muslim scholars, especially Ibn Rushd (Averroës) and Ibn al-Haytham (Alhazen or Alhacen), including in the area of experimental science.[10] "When we examine these Arabian or [Muslim] writers, we cease to feel the same astonishment at the complete contrast between Roger Bacon and his more famous contemporaries. For we find in them the same spirit of scientific inquiry, the same skill in mathematical investigation of all kinds, and the same proneness to practical invention, that are so markedly characteristic of him."[11] This acknowledgement of the shared interest in experiment, mathematical measurement, and invention between Bacon and his forebears and, by implication, Adamson and his own contemporaries, was unusual and short-lived. In his entry on Bacon for the *Dictionary of National Biography* Adamson says merely that "with such aid as was afforded by the increasing knowledge of the Arab writers, [Bacon] devoted himself to acquiring a knowledge of languages, and to experimental researches [*sic*], partly in alchemy, partly in optics."[12] Far from the natural philosopher who "advanced little beyond Alhazen," the version of "Roger Bacon" inscribed in the annals of national biography may have generally relied on "Arab writers" for background reading, but he pursued his experimental research in optics and alchemy (two branches of knowledge specifically associated with earlier Arabic authorities) as an innovative thinker.

The shift evident in both Charles's and Adamson's account to viewing Bacon and his genius ideas as necessarily in contrast to religious ideology crystallized into a theory of historical progress known as the "conflict thesis." This thesis, initially put forth in the extremely popular *History of the Conflict Between Religion and Science* (1875) by John William Draper, and then with even more impact by Andrew Dickson White, cofounder of Cornell University and its first president, in his *A History of the Warfare of Science with Theology in Christendom* (1896), explained the scientific development of Western culture as a result of ideological conflict between the opposing forces of "science" and "religion" and the gradual secularization of Western culture.[13] This idea, that organized religion is hostile to scientific thought and discovery, emerged as a broadly popular notion that fit existing ideas about the necessity of a separation between religion and the public sphere, especially in the United States. White, writing in 1896 about the struggle between "science" and "theology," viewed Bacon as an anachronistic visionary, an experimentalist whose ideas had generated wonderful and useful inventions, and who pursued his inquiries despite the oppressive strictures of the Church. "[In] an age when experimenting was sure to cost a man his reputation, and was likely to cost him his life, he insisted on experimenting, and braved all its risks. Few greater men have lived."[14] The whisper in the historical record that Bacon had been sanctioned for intellectual heterodoxy had become a fact loudly proclaimed, and taken as evidence of Bacon's greatness, his freedom of thought, and his anti-religious tendencies. As Amanda Power put it, "Whether it is intended or not, when Bacon's mind is described as 'modern' there is an inevitable whiff of atheism in the air."[15] Despite Bacon's emphasis on Christian morality expunged from his ideas Bacon still operates as a messianic figure of modernity; his alleged sanction and even imprisonment by Church authorities made him a more modern figure, akin to Galileo.

Even those accounts that denied Bacon's influence on later natural philosophy often credited him with numerous inventions. Hegel, himself a student of medieval philosophy, wrote, "Roger Bacon treated more especially of physics, but remained without influence. He invented gunpowder, mirrors, telescopes, and died in 1294."[16] According to these learned men of the nineteenth century, Bacon had been trapped in some terrible purgatory known as "the Middle Ages," such that his own brilliance went unheeded by his contemporaries.[17] To these writers Bacon was an inventor and visionary, and in some accounts a free-thinker and committed rationalist, who, alone among his peers, articulated something like the modern scientific method. His interest in the past to imagine the future did not translate to these later readers, nor did his commitment to his Order, or his apocalypticism, or his desire for divine illumination and *sapientia*. Yet these later writers were enlightened enough—modern enough—to understand Bacon's ingenuity. Many of his speculative machines had been realized and were a part of their landscape. In addition, Bacon's originality of thought *had* been recognized by those whom they esteemed as the founders of "modern science"—men like Francis Bacon and those of the Royal Society.[18]

The scholarly and popular opinion of Bacon and his role in the history of science established by Whewell and taken up by Charles, Adamson, and White spurred additional interest in Bacon. In the last decades of the nineteenth century, several biographies and studies appeared in England, France, and Germany that further cemented Roger Bacon as both a great man of the past and exemplary of a stage in the history of intellectual development.[19] This interest was largely centered in universities and academic culture—the popularity of "Friar Bacon" had long waned by then. Several decades after Whewell's *History* new printed editions of some of Bacon's previously unpublished works were published in England.[20] In 1897 J. H. Bridges published a new edition of the *Great Work* that was essentially complete. Yet Bacon's Christian faith was increasingly edited out alongside his deep intellectual debts to Arabic authors.

Commemoration and Erasure

The terms of scholarship on Bacon—anticipation, precocity, anachronism—were firmly entrenched by the turn of the twentieth century, and his reputation as one of the great men of science was so well-established that he seemed worthy of greater commemoration. In 1893 Falconer Madan, the Librarian of Oxford University, featured Bacon as an example of the "progressive power" of the scientific tradition at Oxford in his remarks on "Some Interesting Features in the Past History of Science in Oxford" for an open day event with lectures and exhibits at the Museum of Natural History.[21] Madan's comments, which were reported at length in *The Oxford Magazine* the following week, recapitulated the central themes of Whewell and those who followed him: Bacon's contempt for authority, his divergence from his contemporaries, and his insistence on experiment. "His special methods and studies . . . were of such a nature as to . . . shock almost all his contemporaries."[22] Madan summarized *scientia experimentalis* as a set of steps that strongly resembles the scientific method as it is usually taught to schoolchildren: 1) consider the general laws or theories applicable to science; 2) observe the phenomena; 3) apply laws to explain the phenomena; and 4) verify the result.[23] The reporter from *The Oxford Magazine* clearly understood Madan's point when he wrote that Bacon was "scientific to the core."[24] Madan closed his remarks by suggesting that, given Bacon's stature as an intellectual portent, the British Association should commemorate the upcoming six hundredth anniversary of Bacon's death (given as 1294 by Madan); this suggestion was reported (and amplified) in the news coverage.

Perhaps because his suggestion was not taken up by British Association leadership for the 1894 sexcentenary of Bacon's death, Madan remained interested in commemorating Bacon's life and works. Having missed the chance to mark the six hundredth anniversary of Bacon's death, in the intervening years he turned his attention to the upcoming septicentennial of Bacon's birth. Nor was this effort without precedent. Bacon's role as a scientific "harbinger" and evidence for the outsize role that British talent played in the development of science was widespread throughout the United Kingdom in the early twentieth century. For example, Bacon appears as one of the figures on the "science pillar" in St. Anne's Cathedral in Belfast, along with Archimedes (third century B.C.E.), Isaac Newton (1642–1727), and William Thomson, Lord Kelvin (1824–1907). These four figures on one pillar celebrating scientific knowledge together tell a story of *translatio scientiae*, with Britain as the successor to ancient Greece. Bacon appears with alchemical tools, suggesting that, even as material interest in alchemy had declined, he was still identified enough with alchemy and experimentalism so that this depiction would be meaningful.[25] The association with alchemy is somewhat unusual in this period; the emphasis that Adamson, Madan, White, and others place on Bacon as a "modern" scientific thinker meant that his interest in alchemy was more often explained away or remained unmentioned. Bacon's belief in the importance and potential of alchemy to effect material change was an unwelcome reminder of the ways that he was not modern.

Madan was one of the key members of the executive committee launched in March 1914 to raise money to commemorate the seven hundredth anniversary of Bacon's birth. The fund would pay for the publication of his works and for the commission of a statue of Bacon, which was to be installed at the Natural History Museum in Oxford on June 10, 1914.[26] The list of donors to the general fund comprises over one hundred influential and well-respected individuals from Europe and North America. Reflecting Bacon's wide range of interests and topics of study, the donors are notably diverse in terms of their professions and interests, including several Franciscans; arms manufacturer Sir Andrew Noble; economist John Neville Keynes; several classicists; professor and noted aeronautical engineer Bertram Hopkinson; the Librarian of the Patent Office, E. W. Hulme; and William Erasmus Darwin.[27] A number of donors were scholars of the medieval period and a few were archivists and manuscript curators at institutions with significant holdings of Bacon's works. Several, like Pierre Duhem, Karl Sudhoff, and Ludwig Baur, had professional research interests in the history of science and medicine, especially of the medieval period. Duhem was a theoretical physicist, and historian and philosopher of science. His work in the philosophy of science centered on the relationship between theory and experiment; this interest also led him to explore the history of medieval science, especially physics and astral science, which he addressed in several works. Karl Sudhoff, a medievalist at the University of Leipzig, started *Sudhoff's Archiv*, the first scholarly journal in the history of medicine, and Ludwig Baur had in 1914 published an appraisal of the work of English Franciscan natural philosopher Robert Grosseteste (a significant influence on Bacon).[28] Perhaps because of Bacon's philosophical writings, a few prominent philosophers also donated to the cause of his commemoration, including Émile Boutroux and Henri Bergson. Others were leaders in higher education and politics, such as George, Lord Curzon, the former viceroy of India and at that time chancellor of Oxford; Arthur Balfour, former prime minister of the United Kingdom (1902–5) and future foreign secretary; and Nicholas Murray Butler, president of Columbia University. A large number of scientists gave to the fund, perhaps because the Executive Committee, charged with planning and fundraising, was headed by Sir Archibald Geikie, the eminent geologist and former president of the Royal Society, and included other esteemed men of science, such as Sir William Crookes, recipient of the Davy and Copley medals of the Royal Society and its president, and Sir William Osler, a Fellow of the Royal Society and famous physician. The scientists who donated were also highly accomplished and recognized for their work: many were fellows of the Royal Society, and others, like Sir Alexander Miers and Lucien Poincaré, occupied positions of institutional power and authority.[29]

The seventh centenary of Bacon's birth and its commemoration at Oxford received international attention. The installation of the statue of Bacon at the Natural History Museum featured remarks from Geikie and Lord Curzon, as well as from the university's public orator, A. D. Godley, and was attended by donors to the General Fund and members of the press. Newspapers and journals worldwide, including *The Times Literary Supplement*, *The Saturday Review*, *Nature*, *L'Osservatore Romano*, *Revue des deux Mondes*, and *The British Medical Journal*, ran stories about the event and about Bacon's contributions to science.[30] Catholic publications in the Anglophone world also remembered Bacon and his achievements, but usually acknowledged the role of Christianity in Bacon's life.[31] A full issue of *The Open Court*, a monthly journal of science and religion, dedicated to Bacon and his contributions to science appeared in August 1914, and a scholarly volume of essays tied to the Oxford celebration appeared later in the year.[32]

The commemorative spirit burned in the United States as well. In addition to contributing to the general fund, Nicholas Murray Butler, the president of Columbia University, commissioned a play from English professor John Erskine, founder of the General Honors course that eventually became the "Great Books" component of Columbia's Core Curriculum. *A Pageant of the XIIIth Century*, written in iambic pentameter, was scheduled to be performed in November 1914 but the performance was cancelled due to World War I.[33] Erskine's pageant, to have taken place at Columbia University, reflects the view of Roger Bacon, the Middle Ages, and the history of science that prevailed in academia and more broadly. Written within a university context, it would have been performed for students, other faculty, community members, and their families. In the play Erskine omitted Bacon's reliance on Arabic authors and presented medieval Arabic science as tethered to Greek science and organized around translation: The only significant character from Islamdom that Erskine included is philosopher Ibn Rushd (called by his Latin name, Averroes), who is credited with summarizing Greek philosophy for

Latin Christians and "[bringing] to the Christian nations an understanding of Aristotle and of science."[34] The play also includes a brief scene set in "The House of Wisdom," the legendary site of Greco-Arabic translation in ninth-century Baghdad.[35] The *bayt al-Ḥikma*, or "House of Wisdom," as a specific, centralized location in Baghdad established for the translation of Greek scientific texts into Arabic in the early ninth century was often-repeated in twentieth-century scholarship, and it conveniently supports the assumption that intellectual endeavor in Islamdom in this period was limited to the translation and copying of Greek scientific texts. However, more recent scholarship has reestablished *bayt al-Ḥikma* as an Arabic translation of the Sasanian Persian term for "library" (in Sasanian context, one specifically for historical material) and has contextualized this institution as one part of an extensive administrative bureaucracy.[36] Aside from these two glimpses of Arabic intellectual culture, Erskine reduced medieval Islamdom and its varied, complex cultures to the broadest stereotypes of Orientalism—the crusades and a "Saracen dancing girl" at the court of Frederick II.[37] This one-dimensional portrait of the Islamicate world went along with the erasure of Arabic authorities on natural philosophy; how could an incurious and decadent culture produce anything resembling scientific thinking?

The versions of Roger Bacon and the Middle Ages presented in the Oxford and Columbia commemorations owed a great deal to Whewell, but they were also indicative of early twentieth-century currents. Contrary to Whewell's goal to unify Christian morals and scientific knowledge, positivist historians, including George Sarton, the founder of the discipline of the history of science, viewed Bacon, "free thinker and persecuted by the Church," as evidence that science had won out over superstition.[38] At Columbia, Nicholas Murray Butler and John Erskine were both heavily involved in shaping the canon of what became known as "Western thought" in the early twentieth century. Butler instituted what became the Core Curriculum at Columbia University and the Contemporary Civilization course and recruited both the medieval period and Bacon in the service of the Western Civilization curriculum.[39] And Erskine designed the General Honors course—what became known as the "Great Books" course—at Columbia, which was the literature counterpart to Butler's history course. Just as Bacon saw a need to reform the university curriculum to strengthen Christendom, Erskine saw a need to reform the curriculum to strengthen American civil society by promoting an idea of "Western civilization" based on great "Anglo-Saxon" [*sic*] writers. This focus on what is often called the "Western canon" was to give students from immigrant backgrounds—particularly Jewish students—the knowledge they would need to assimilate into American culture.[40]

By the seventh centenary of Bacon's birth in 1914 his engagement with and reliance on Arabic authors like Ibn al-Haytham and al-Kindī was no longer part of the narrative about his genius. None of the speeches and articles recapitulating Bacon's importance to the history of science from the septicentennial of his birth mention the influence of Arabic texts and ideas. To do so would be to acknowledge that Arabic authors did much more than translate Greek texts, and that Bacon was not the sole inventor of a new method of knowledge acquisition. His role as the inventor of a method of natural inquiry that closely resembled the methods of modern science was the reason for his commemoration, as the press reports and speakers' remarks made clear. This invention, to Whewell and those who followed, demonstrated the supposed inevitability of the development of modern science in Europe. The celebration of Bacon as a modern or proto-modern figure in a period whose defining characteristic is that it is *not* modern requires Bacon to represent the worldview on which scientific modernity rests—a rejection of authority and superstition in favor of direct experimentation and a progressive view of science. This misreading of Bacon and his context is in service of the view that scientific thinking—and thus modernity—are intrinsic to Christian Europe, while people of other religions and in other places remain trapped in "the Middle Ages" by religious faith, submission to authority, and lack of curiosity. In this way "the Middle Ages" had become, in addition to an invented historical period, a developmental stage in human progress. Bacon's subjectivity as an educated Christian Englishman from an elite family is central to this view. Because his interests in dragon flesh, the apocalypse, astral prediction, and attaining *sapientia* undermine the portrait of "Roger Bacon, secular rationalist," they were downplayed or ignored; just as his Arabic influences and his interests in a pseudo-Aristotelian text were erased or overlooked.

History and Historiography

The priorities of Whewell and his contemporaries set the terms of scholarship on Bacon throughout the twentieth century in the history of science, which has remained focused on questions about Bacon's precocity, especially in the areas of experimentalism, invention of new technologies, and the mathematical foundation of natural science. Even those attempts to correct the hagiographical view of Bacon within the history of science remain bound by the terms set by nineteenth-century writers: invention, precocity, priority.[41] In the early twentieth century, Pierre Duhem—one of the subscribers to the commemorative fund in 1914—submitted that Bacon's argument for *scientia experimentalis* was a praise of observation in the methods of natural philosophy, while in 1914 Lynn Thorndike Jr. argued that Bacon was not exceptional and had been taken out of context, and that his sources had been overlooked or ignored.[42] Thorndike developed this view in later scholarship, writing of Bacon a decade later: "Roger Bacon has hitherto been studied too much in isolation. He has been regarded as an exceptional individual; his environment has been estimated at his own valuation of it or according to some preconceived idea of his age; and his writings have not been studied in relation to those of his predecessors and contemporaries. Thought of as a precursor of modern science, he has been read to find germs of modern ideas rather than scrutinized with a view to discovering his sources."[43]

But Thorndike was only one voice among many. George Sarton, writing around the same time, argued that Thorndike's appraisal was too harsh, and maintained Bacon's originality and priority. "He was not himself an experimenter any more than he was a mathematician, but he clearly saw—better than anybody else in his time—that without experimentation and without mathematics, natural philosophy is very soon reduced to verbiage."[44] More broadly, debates about Roger Bacon's influence are also always about the importance of the Middle Ages—especially the Latin Middle Ages—to the development of modern science, and thus to the history of science as well. Sarton characterized Bacon as less systematic in his thinking than some of his peers, like St. Thomas Aquinas, but more original. "And the best proof of this is that he was a true harbinger of modern civilization, while St. Thomas (not to speak of the others) was destined to become the most perfect symbol of mediaevalism."[45]

How important is that context—learned Latin culture—to understanding how modern science developed? This question has occupied historians of medieval science since the earliest days of the profession in the twentieth century. Pierre Duhem, writing at the turn of the twentieth century, argued in favor of continuity of scientific thought between the Parisian masters of the fourteenth century and early modern natural philosophers.[46] Since Duhem, scholars have disputed whether continuity or discontinuity (or both) characterizes the relationship between these historical periods. Lynn Thorndike, writing just a few years later, explicitly linked the natural magic of the Latin Middle Ages with later experimental science in his eight-volume work *The History of Magic and Experimental Science*. Yet some early modernists were not convinced, arguing instead that early modern science represents a moment of rupture with previous beliefs and practices.[47] In turn, some medieval historians argued that science in the seventeenth century had inherited from the medieval period a robust and highly rationalized methodology.[48] Others argued for general discontinuity but continuity at the level of specific sciences, or first in favor of discontinuity, and then, later, in favor of direct influence.[49]

Bacon's visionary nature and originality of thought continued to be a scholarly focus. Bacon became linked with the mysterious Voynich cipher manuscript in the 1920s when William Newbold, a professor at the University of Pennsylvania, claimed, without evidence, that Bacon was the author and had invented the telescope and the microscope.[50] Newbold's claims were initially believed then publicly demolished after his death, but the link between Bacon and esoteric invention persisted. A few decades later, Alistair Crombie argued that, although Bacon's ideas about *scientia experimentalis* were not sui generis, he significantly contributed to the development of modern science by his application of Grosseteste's ideas on *experimentum*.[51] If Bacon owes anything to earlier scholars, in Crombie's account, it is to the fellow Franciscan and Englishman Robert Grosseteste, rather than to Arabic authorities.

By the middle of the twentieth century Bacon was no longer spoken of in quite the same hagiographic terms. Biographies by Stewart Easton and Theodore Crowley did a great deal to place Bacon within his context.[52] But by this point both the historical periodization in which Bacon was embedded and the narrative of scientific progress first established by Whewell were fully entrenched as received wisdom. Both these ideas—that there was a long period of intellectual stagnation in Christian Europe between antiquity and the Renaissance and that modern experimental science only emerged in Christian Europe—form the ground for Thomas Kuhn's argument for paradigm-shifting discoveries in the period immediately following the long middle. They also form the ground for his assertion that "only the civilizations descended from Hellenic Greece have possessed more than the most rudimentary science. The bulk of scientific knowledge is a product of Europe in the last four centuries."[53] It would be more accurate to say Kuhn's assertion and the confidence behind it are a product of a European historical construct in the last four centuries. Kuhn's statement about the development of science is akin to looking at a manicured garden, with a boxwood maze, cultivated flower beds, and topiaries ruthlessly pruned into fantastical shapes and marveling that Nature should have arranged it thus.

The historiography of Roger Bacon that has remained in place since Whewell promotes a narrative of medieval natural philosophy and technology that affirms later values, such as a narrow definition of experimentalism and a belief in the importance of innovation, and that colonizes the past in the service of the ideological constructs of "the West" and "modern science." Roger Bacon in the historiography of science reveals the importance of periodization, and of the medieval period, to the idea of modern science. Discussions about Bacon's originality or his precocity rely on a context of an imagined "Middle Ages," one without any of the complexity, texture, or specificity of reality. The origins of this historical epoch and the negative valuation attached to it are bound up with Christian nationalism and European imperialism. Scholars of the period are aware of the origins of "the Middle Ages" and the ways that many of the tools and resources for studying the medieval past were created in the service of nationalism, imperialism, and white supremacy, and how medievalism remains vital to contemporary white supremacist, Islamophobic, and anti-Semitic ideologies.[54] The historiography of Roger Bacon is another example of the role that historical periodization and "the Middle Ages" played in supporting and promoting scientific racism; in this case as a historical example used to justify the ideology of Western intellectual superiority.

More surprisingly, the historiography of Roger Bacon makes visible the extent to which many of the tools for inventing modern science are medieval tools. The regime of historicity and universalizing Christian typology that emerged in late antiquity and was systematized in the twelfth and thirteenth centuries formed Whewell's schema for the development of scientific thinking over time. This frame, of potentiation and progress, with "winners" and "losers" along the way, remains in place in the historiography of science.[55] As a result Bacon's contemporaries, as well as everyone else living in the medieval period, are denied the same status as their modern counterparts. Bacon alone has been granted that status. The result is a narrative of scientific development that uses Christian historicity to fashion an argument about modernity, with eurocentrism and Christian supremacy intact at its core.

Combining an understanding of medieval ideas about periodization (how everything from before Christianity became periodized as part of it) with how the Middle Ages has been periodized as anti-modern yields new insight into the historiography of science and medieval culture. Bacon's inspirations for both past and future cohere around a set of ideas that privilege hands-on knowledge, first-hand observation, utility, and textual knowledge from Arabic sources, literary and historical narratives, and scholastic treatises on *argumentum* and natural philosophy. He looks backward to look forward, blending accounts of the past with new knowledge from his contemporaries. His account of *scientia experimentalis* and the incredible possibilities it affords offers a counterpoint to the contemporary narrative of technological development, which sees the imagination of new technologies and their uses as something that is relentlessly progressive and future-oriented and views the inventions of the past as simple or obsolete.

Conclusion

Bacon's Ghosts

> By the latter part of the thirteenth century, Europe was only beginning to rouse from its intellectual stupor. Alhazen's work was translated into Latin, and it had a great effect on the writings of Robert Grosseteste (1175–1253), Bishop of Lincoln, and on the Polish mathematician Vitello (or Witelo), both of whom were influential in rekindling the study of Optics. Their works were known to the Franciscan Roger Bacon (1215–1294), who is considered by many to be the first scientist in the modern sense.
>
> —Eugene Hecht, *Optics*, 2008

Our story began with the intellectual biography of an individual and a particular set of ideas. In Oxford and Paris Bacon was part of an intellectual current of men interested in particular phenomena and *experimenta*, and in larger epistemic questions about how to understand and trust knowledge from the senses. In addition to his teachers, colleagues, and students, Bacon also drew on knowledge that was not part of the academic curriculum to articulate and explain *scientia experimentalis*. From texts on history and rulership he drew concrete examples about the technologies and outcomes that *scientia experimentalis* could foster; in this way, his theory—lauded for centuries as new—was strongly grounded in a retrospective and aspirational view of the past, rather than offering a completely new vision of the future. And from texts written in Arabic and newly available to Bacon in Latin translation he borrowed, synthesized, and Christianized knowledge about optics, experimentation, alchemy, and astral science, as well as other branches of knowledge. His method of knowledge-acquisition and his plans for curricular reform are inextricable from his Christian belief and Christian identity, and his desire to strengthen Christendom morally, politically, and militarily. We ended up with "Roger Bacon," a figure who lived during a period of intellectual stagnation, out of step with his moment; who was in open conflict with religious orthodoxy; who worked alone to invent a new method for understanding natural laws and phenomena; and who was misunderstood and punished for his innovation. This stunning transformation was effected over centuries through partial circulation of Bacon's work, especially related to *scientia experimentalis* and adjunct subjects; creative translations of Bacon's work and misattributions to his authorship; passive effacement and active suppression of Bacon's reliance on Arabic sources; the invention of "the Middle Ages" as a historical period defined as unmodern and backward; and the ideological imperatives of Christian supremacy and Western intellectual superiority.

Over centuries, Bacon was constituted as an exceptional figure because of his exemplarity: a modern figure in an unmodern time. According to the logic of typology, Bacon typifies modern scientific thinking, albeit incompletely or imperfectly (for example, his Christian fundamentalism and belief in alchemy make him imperfectly modern). An exceptional figure in an exceptionally bad time, he functions as the prophetic figure—the "harbinger of modern civilization" in the words of George Sarton—whose presence in the Middle Ages is taken as proof of the uniquely European and Christian origins of modern science.[1] But exceptionalism is configured differently in a period when most of the historical actors about whom we know are exceptional, either because they left a significant trace in the historical record or because of their status as literate or elite. The record that has survived to us from the medieval period is contingent and accidental, and it privileges certain minorities: elites, literate people, and individuals who crossed paths with institutional authority. Seeing Bacon as exceptional for his time and exemplary of a later worldview makes it difficult to see him exemplary of his own time and place. He may have been exceptionally productive, but he was also exceptional because of his status (from a baronial family), his independent wealth, and the company he kept. By examining Bacon over the longue durée and including his afterlives it becomes clear that exemplarity and exceptionalism often work to uphold unstated ideological positions.[2] However, exceptionalism and exemplarity can be reconfigured in historiography and mobilized differently.

Tracing the invented histories of Roger Bacon across centuries and comparing the thirteenth-century portrait with the later versions reveals that there is nothing self-evident about historical periodization or the historical narrative of scientific development. The medieval period as a historical era or a stage of human development is a later invention, a necessary foil to modernity. We may be stuck with the Middle Ages due to inertia, but it is worth remembering the origins of the ancient-medieval-modern periodization and the ideology it sustains. As I have shown, creating and sustaining this historical periodization and the narrative of scientific progress has required plenty of historical accident but also purposeful invention and intervention from historical actors.

At the same time, the long view also makes possible some moments of reverberation and resonance between the thirteenth century and later. Certainly, plenty of later writers found in Bacon a thinker whose ideas seemed familiar, and like their own. The use of natural knowledge for political domination as well as personal moral improvement aligned with the views of men like Dee and the Fellows of the Royal Society. Bacon's argument for the necessity of rationality and experience working together to make reliable knowledge about the world echoed in the work of Francis Bacon and William Whewell. The early twenty-first century likewise offers possible resonances with Bacon's moment; he wrote from a position of deep concern about the viability of the future and with the conviction that the university curriculum was not sufficient to meet the political and epistemic challenges of the moment.

Into the twenty-first century, Bacon has been resurrected as a subject of serious scholarship by historians of science. Over the past thirty years Bacon's own ideas, texts, and context have come into greater focus, alongside a more expansive view of natural knowledge in the long medieval period.[3] The limitations of the historical concept of what used to be called "the Scientific Revolution" have been thoroughly discussed; likewise, the essential wrongness of the conflict thesis.[4] Scholars of the early modern period have demonstrated the centrality of non-European actors, ideas, and places to the development of modern science, and have identified multiple macro-regional or even global networks of knowledge and exchange.[5] Experiment and experimentation have been identified in contexts beyond the scientific academies and universities of early modern Europe, and practiced by non-elite actors in multiple contexts.[6]

Although Roger Bacon and his ideas are now viewed in their own context and on their own terms in the history of science, the idea of Bacon and his period, as Amanda Power put it, "as a 'stage' in a long, triumphant history," remains intact and current within the natural sciences.[7] Introductory-level college textbooks in the exact sciences, especially in physics, astronomy, cosmology, and optics, continue to rehearse Whewell's narrative of scientific development. A major textbook on optical science opens with a brief overview of the history of optics that begins with the book of Exodus in the Hebrew Bible before moving ahead to the ancient Greeks. Then, "after the fall of the Western Roman Empire (475 A.D. [*sic*]), which roughly marks the start of the Dark Ages, little or no scientific progress was made in Europe for a great while."[8] In this account the Christianity remains intact, signaled by the use of "A.D." for "*anno domini*" or "the year of our Lord," dated from the Christian Incarnation. Other textbooks in physics and astronomy repeat the same narrative without the Christian foundation, either referring to the "Dark Ages" as a time when scientific knowledge stalled or simply skipping from the ancient Greeks to the early modern period, like Kuhn.[9] In some instances, the development of scientific knowledge is explicitly addressed as a pedagogical tool that also reflects the development of humanity itself. "It is customary to start learning astronomy from a historical perspective. This is a natural way to learn about the universe; it permits our personal growth in knowledge to echo humanity's growth in knowledge, starting with relatively nearby and familiar objects, and then moving outward."[10] In this formulation, earlier historical periods are conceived as intellectually immature. The implicit promise to the twenty-first century reader is that the student will be able to outgrow their immaturity by learning more advanced astronomy than was available to the ancient or medieval expert. The student inhabits a different time period, existing in an allochronic relationship with people from long ago. But the student's ignorance confers coevality with the people from long ago, at least until the student learns enough to know better. In textbooks across the sciences, the history of science is presented as a natural product of Christian Europe, and as part of the history of Western intellectual superiority.

In the popular culture of the late twentieth and early twenty-first centuries, Bacon has often retained his association with intellectual rebellion and esoterica. He is one of the models of committed empiricist and Franciscan William of Baskerville in Umberto Eco's medieval mystery *The Name of the Rose* and of the disgraced Maester Qyburn in George R. R. Martin's historical fantasy series *A Song of Ice and Fire*.[11] Bacon is often associated with historical fantasy or science fiction, a reflection of the identification of the medieval period with the esoteric or unsettling and the link between Bacon and anachronism.[12]

Exploring Roger Bacon and the wider context of medieval science makes it possible to understand him as indicative of a shared interest in experiment, rather than a lone figure. Instead of being considered exceptional for his time, a heterodox thinker who anticipated modernity, Bacon is exemplary of his thirteenth-century milieu, characterized by interest in *experimenta*, measurement, and instruments. One way to say *we are like him* is to understand the unmodern period—in thirteenth century learned Latin circles, or in eleventh-century Egypt, and many other places besides—as capable of producing individuals and groups who devised systematic ways to understand natural phenomena and offer a predictive model, and therefore to reconsider what is modern about modern science. If experimental thinking emerged in places and times that are definitively marked as unmodern, perhaps it makes more sense to consider experimental thinking as akin to agriculture or the wheel, something invented by people in multiple time periods, places, and contexts, and then adopted, or not, for all kinds of reasons.[13] Saying *we are like Roger Bacon* allows the possibility of asking different questions, ones that begin from a recognition of equal humanity between historian and historical actor, and a willingness to be taught by the past, rather than condescending to it.

Notes

INTRODUCTION

Notes to epigraphs. "Nam licet naturae potens sit et mirabilis, tamen ars utens natura pro instrument potentior est virtute naturali, sicut videmus in multis." Roger Bacon, *Epistola de secretis operibus artis et naturae, et de nullitate magiae*, in *Fr. Rogeri Bacon opera quædam hactenus inedita*, ed. J. S. Brewer (Longman and Green, 1859), 523–51, 523. Unless otherwise noted, all translations are my own. William Whewell, *History of the Inductive Sciences: From the Earliest to the Present Time*, 2 vols., 3rd ed. (J. W. Parker, 1857 [1837]), 1:3. Thomas S. Kuhn, *The Structure of Scientific Revolutions*, 4th ed. (University of Chicago Press, 2012 [1967]), 167.

1. "Impetu inaestimabili"; "terrores infiniti," Bacon, *Epistola*, 533, 534. This text has not always been accepted as a genuine work of Bacon's because the work, as it has come down to us, contains material in the later chapters that some scholars have deemed too different in tone and substance from Bacon's other works. However, more recent scholarship suggests that the first eight chapters of this work closely follow Bacon's earlier works and are likely genuine, and there is additional compelling evidence to suggest that the final three chapters, which differ from the preceding eight, are also by Roger Bacon. See Dorothea W. Singer, "Alchemical Writings Attributed to Roger Bacon," *Speculum* 7, no. 1 (1932): 80–86; William R. Newman, "The Philosophers' Egg: Theory and Practice in the Alchemy of Roger Bacon," *Micrologus* 3 (1995): 75–101; Meagan S. Allen, *Roger Bacon and the Incorruptible Human, 1220–1292: Alchemy, Pharmacology and the Desire to Prolong Life*, Palgrave Studies in Medieval and Early Modern Medicine (Palgrave Macmillan, 2023), 233. Allen is also preparing a critical edition and translation of this work. Many of Bacon's works do not exist in critical editions, in part due to his habits of writing multiple drafts and repurposing his work under multiple titles, which make manuscripts of his work difficult to organize. See A. G. Little, "Roger Bacon's Works with References to the MSS. and Printed Editions," in *Roger Bacon: Essays*, ed. A. G. Little (Clarendon Press, 1914), 375–425; Jeremiah Hackett, "Roger Bacon: His Life, Career, and Works," in *Roger Bacon and the Sciences: Commemorative Essays*, ed. Jeremiah Hackett (Brill, 1997), 9–23; Amanda Power, *Roger Bacon and the Defence of Christendom* (Cambridge University Press, 2012), 11–12. The following editions, however imperfect, are in standard use: Roger Bacon, *Opus majus*, ed. John Henry Bridges, 3 vols. (Clarendon Press, 1897–1900); Roger Bacon, *Opus minus*, in *Fr. Rogeri Bacon opera*, 313–89; Roger Bacon, *Opus tertium*, in *Fr. Rogeri Bacon opera*, 3–310; Roger Bacon, *Opus tertium*, in *Part of the Opus Tertium of Roger Bacon, Including a Fragment Now Printed for the First Time*, ed. A. G. Little (Aberdeen University Press, 1912); Roger Bacon, *Un fragment inédit de l'Opus tertium de Roger Bacon précédé d'une étude sur ce fragment*, ed. Pierre Duhem (Quarrachi, 1909); and *Opus*

tertium, ed. and trans. Nikolaus Egel (Felix Meiner, 2020). The standard English translation of *Opus majus*, by Robert Belle Burke, 2 vols. (University of Pennsylvania Press, 1928) can be misleading; he offers "science" for *scientia* and "experimental" for *experimentalis* (both of which I translate differently below). More recently (and only partially), see Ruggero Bacone, *Filosofia, scienza, teologia dall'Opus Maius*, trans. Valeria Sorge and Fabio Seller (Armando, 2010); *La scienza sperimentale: Lettera a Clemente IV—La scienza sperimentale—I segreti dell'arte e della natura*, trans. Francesco Bottin (Rusconi, 1990). For Bacon's letter to Pope Clement IV, see also Roger Bacon "Epistola Fratris Rogeri Baconi," *Lettera a Clemente IV*, ed. and trans. P. Efrem Bettoni (Biblioteca Francescana Provinciale, 1964).

2. Note on names: Throughout this book "Bacon" only refers to Roger Bacon; Francis Bacon is always referred to with his cognomen.

3. I keep Bacon's term, *scientia experimentalis*, throughout.

4. On the shift from science as a type of knowledge to a method of knowledge-production, see Henry M. Cowles, *The Scientific Method: An Evolution of Thinking from Darwin to Dewey* (Harvard University Press, 2020).

5. E.g., Edmund Wrigglesworth, "Roger Bacon," *Universe, an Illustrated Catholic Weekly Newspaper*, June 5, 1914; Cecil Underwood, "Roger Bacon," *Ave Maria*, June 20, 1914, 769–75; Ernst Dühring, "The Two Bacons," trans. Lydia G. Robinson, *Open Court* 28, no. 8 (August 1914): 468–85.

6. Amanda Power outlines this trajectory in her article "A Mirror for Every Age: The Reputation of Roger Bacon," *English Historical Review* 121, no. 492 (2006): 657–92. Her article introduced me to some of the archival sources and early modern editions on which I rely for this study.

7. Whewell, *History of the Inductive Sciences*, 1:254.

8. Johannes Fabian, *Time and the Other: How Anthropology Makes Its Object* (Columbia University Press, 2014 [1983]). The terms "schizogenic," "coeval," and "allochronic" are Fabian's.

9. John Dagenais and Margaret Rich Greer in "Decolonizing the Middle Ages: Introduction," in "Decolonizing the Middle Ages," special issue, *Journal of Medieval and Early Modern Studies* 30, no. 3 (2000): 431–48; Catherine Brown, "In the Middle," *Journal of Medieval and Early Modern Studies* 30, no. 3 (2000): 547–74; Jeffrey Jerome Cohen, ed., *The Postcolonial Middle Ages* (Palgrave, 2000); Patricia Clare Ingham and Michelle R. Warren, eds., *Postcolonial Moves: Medieval through Modern* (Palgrave Macmillan, 2003); Nadia Altschul, "Postcolonialism and the Study of the Middle Ages," *History Compass* 6, no. 2 (2008): 588–606; Dipesh Chakrabarty, *Provincializing Europe: Postcolonial Thought and Historical Difference* (Princeton University Press, 2007 [2000]); Kathleen Davis, *Periodization and Sovereignty: How Ideas of Feudalism and Secularization Govern the Politics of Time* (University of Pennsylvania Press, 2008); Dipesh Chakrabarty, "Historicism and Its Supplements: A Note on a Predicament Shared by Medieval and Postcolonial Studies," in *Medievalisms in the Postcolonial World: The Idea of "the Middle Ages" Outside Europe*, ed. Kathleen Davis and Nadia Altschul (Johns Hopkins University Press, 2009), 109–19; Carol Symes, "When We Talk about Modernity," *American Historical Review* 116, no. 3 (June 2011): 715–26.

10. Fabian, *Time and the Other*, 3. Others have also critiqued this aspect of Fabian's account; for example, see Davis, *Periodization and Sovereignty*, 2, 12–13.

11. Whewell refers to the Middle Ages as "the Stationary Period" in the title of Book IV, Chapter 4, see Table of Contents, *History of the Inductive Sciences*.

12. Power, "A Mirror for Every Age," 677–79.

13. "Regime of historicity" refers to the (apparently) modern temporal relationship between past, present, and future, taken from François Hartog, *Régimes d'historicité: Présentisme et expériences du temps* (Seuil, 2003), published in English as *Regimes of Historicity: Presentism and Experiences of Time*, trans. Saskia Brown (Columbia University Press, 2015). I have taken the term from Hartog, but disagree with his analysis, which articulates the regime of historicity before the French Revolution as focused on the biographies of great men. On "Sattelzeit," the period between 1750–1850, see Otto Bruner, Werner Conze, and Reinhart Koselleck, eds., introduction to *Geschichtliche Grundbegriffe: Historisches Lexikon zur politisch-sozialen Sprache in Deutschland* (Klett-Cotta, 1972), 1:xiii–xxvii. See also Almuth Ebke and Christoph Haack, "Periodization and Modernity: An Introduction," in "Periodization and Modernity," ed. Almuth Ebke and Christoph Haack, special issue, *History of European Ideas* 51, no. 2 (2025): 307–20; and Fernando Esposito, "Historicizing Historicism: Reinhard Koselleck and the Periodization of Modernity," in "Periodization and Modernity," ed. Almuth Ebke and Christoph Haack, special issue, *History of European Ideas* 51, no. 2 (2025): 321–36.

14. Magda Teter, *Christian Supremacy: Reckoning with the Roots of Antisemitism and Racism* (Princeton University Press, 2023).

15. Kathleen Biddick, *The Typological Imaginary: Circumcision, Technology, History* (University of Pennsylvania Press, 2003), 1–6. On how meaning arises from narrativity, see Hayden White, "The Value of Narrativity in the Representation of Reality," in *The Content of the Form: Narrative Discourse and Historical Representation* (Johns Hopkins University Press, 1987), 1–25.

16. For how Bacon has been defined in the history of science since Whewell, see Pierre Duhem, *Le système du monde: Histoire des doctrines cosmologiques de Platon à Copernic*, 10 vols. (Hermann, 1913–59), 3:442; George Sarton, *Introduction to the History of Science*, 3 vols. (Johns Hopkins University Press, 1927–47), 2:951–65; A. G. Molland, "Roger Bacon as Magician," *Traditio* 30 (1974): 445–60; Alistair C. Crombie, *Robert Grosseteste and the Origins of Experimental Science, 1100–1700*, 2 vols. (Clarendon Press, 1953), 1:156; Alexandre Koyré, "The Origins of Modern Science: A New Interpretation," *Diogenes* 4, no. 16 (1955): 1–22; David C. Lindberg, "Roger Bacon's Theory of the Rainbow: Progress or Regress?," *Isis* 57, no. 2 (1966): 235–48; David C. Lindberg, "Lines of Influence in Thirteenth-Century Optics: Bacon, Witelo, and Pecham," *Speculum* 46, no. 1 (1971): 66–83; David C. Lindberg, "Science as Handmaiden: Roger Bacon and the Patristic Tradition," *Isis* 78, no. 4 (1987): 518–36; Jeremiah Hackett, "*Ego Expertus Sum*: Roger Bacon's Science and the Origins of Empiricism," in *Expertus sum: L'expérience par les sens dans la philosophie naturelle médiévale*, ed. Thomas Bénatouïl and Isabelle Draelants, Micrologus 40 (SISMEL Edizioni del Galluzzo, 2011), 145–73; Brian Clegg, *The First Scientist: A Life of Roger Bacon* (Carroll and Graf, 2003); Power, "A Mirror for Every Age," 658. Although arguing against the prevailing idea in 1914 that Bacon had been somehow ahead of his time, scholars like Lynn Thorndike Jr. and A. G. Little were evaluating Bacon according to terms set by Whewell. See Lynn Thorndike, *A History of Magic and Experimental Science*, 8 vols. (Columbia University Press, 1923–58), 2:618–19; Lynn Thorndike, "Roger Bacon and Experimental Method in the Middle Ages," *Philosophical Review* 23, no. 3 (1914): 271–98; A. G. Little, "On Roger Bacon's Life and Works," in *Roger Bacon: Essays*, ed. A. G. Little (Clarendon Press, 1914), 3–31.

17. I briefly outline this legend and its repeated appearance in the historiography about Bacon in further detail below, and an in-depth consideration of the sources for Bacon's supposed condemnation and the treatment of this vignette in later historiography can be found in more detail in Amanda Power, "Seeking Remedies for Great Danger: Contemporary

Appraisals of Roger Bacon's Expertise," in *Knowledge, Discipline, and Power in the Middle Ages: Essays in Honour of David Luscombe*, ed. Joseph Canning, Edmund King and Martial Staub (Brill, 2011), 63–78; and Power, *Roger Bacon and the Defence of Christendom*, 15–26.

18. Andrew Dickson White, *A History of the Warfare of Science with Theology in Christendom*, 2 vols. (D. Appleton and Co., 1896), 1:247.

19. "Cum vero jam monstravi proprietatem hujus scientiae secundem se, jam cuilibet patet quod utilissima est haec scientia post morale, et primo ipsi theologiae absolutae propter sensum literalem et spiritualem, in quibus ipsa consistit" (In truth, I demonstrated this property, that according to this branch of knowledge, which is, next to moral philosophy the most useful, will present the literal truth of Scripture, so that through suitable adaptations the spiritual sense may be derived): Bacon, *Opus maius* 6.12, ed. Bridges, 2:219–20. See also Jeremiah Hackett, "Roger Bacon on *Scientia Experimentalis*," in *Roger Bacon and the Sciences: Commemorative Essays*, ed. Jeremiah Hackett (Brill, 1997), 277–316, 310; Power, *Roger Bacon and the Defence of Christendom*, 152–56.

20. Bacon, *Opus majus* 6.12, ed. Bridges, 2:211–12; F. A. Gasquet, "An Unpublished Fragment of a Work by Roger Bacon," *English Historical Review* 12, no. 47 (1897): 494–517, at 498. See also Zachary A. Matus, *Franciscans and the Elixir of Life: Religion and Science in the Later Middle Ages* (University of Pennsylvania Press, 2017), 50–80.

21. Dagenais and Greer, "Decolonizing the Middle Ages: Introduction"; Michelle R. Warren, *Creole Medievalism: Colonial France and Joseph Bédier's Middle Ages* (University of Minnesota Press, 2011).

22. See Wouter J. Hanegraaff, *Esotericism and the Academy: Rejected Knowledge in Western Culture* (Cambridge University Press, 2012) on this point.

23. On alchemy and alchemical medicine, see William R. Newman, "The Alchemy of Roger Bacon and the *Tres epistolae* Attributed to Him," in *Comprendre et maîtriser la nature au Moyen Age: Mélanges d'histoire des sciences offerts à Guy Beaujouan*, ed. Danielle Jacquart (Droz, 1994); William R. Newman, "The Philosophers' Egg"; William R. Newman, "An Overview of Roger Bacon's Alchemy," in *Roger Bacon and the Sciences*, ed. Hackett, 317–36; William R. Newman, "Roger Bacon and Secrecy: Alchemical Techniques of Deception and the *Doctor Mirabilis*," in *The Sciences of Roger Bacon's* Opus maius, ed. Meagan S. Allen (Routledge, 2025), 165–83; Faye Getz, "To Prolong Life and Promote Health: Baconian Alchemy and Pharmacy in the English Learned Tradition," in *Health, Disease, and Healing in Medieval Culture*, ed. Sheila D. Campbell, Bert S. Hall, and David N. Klausner (St. Martin's Press, 1992), 141–51; Faye Getz, "Roger Bacon and Medicine: The Paradox of the Forbidden Fruit and Long Life," in *Roger Bacon and the Sciences*, ed. Hackett, 337–64; Agostino Paravicini Bagliani, *Medicina e scienze della natura alla corte dei papi nel Duecento* (Centro Italiano di Studi Sull'alto Medioevo, 1991); Agostino Paravicini Bagliani, "Ruggero Bacone e l'alchimia di lunga vita: Riflessioni su il testi," in *Alchimia e medicina nel medioevo*, ed. Chiara Crisciani and Agostino Paravicini Bagliani (SISMEL Edizioni del Galuzzo, 2003); Leah DeVun, *Prophecy, Alchemy, and the End of Time: John of Rupescissa in the Late Middle Ages* (Columbia University Press, 2009), 80–89; Matus, *Franciscans and the Elixir of Life*; Meagan S. Allen, "Alchemy and Pharmacology in the *Opus maius*," in *The Sciences of Roger Bacon's* Opus maius, ed. Meagan S. Allen (Routledge, 2025), 147–64; Allen, *Roger Bacon and the Incorruptible Human*. On secrets of nature and books of secrets in this period, see Steven J. Williams, *The Secret of Secrets: The Scholarly Career of a Pseudo-Aristotelian Text in the Latin Middle Ages* (University of Michigan Press, 2003); William Eamon, *Science and the Secrets of Nature: Books of Secrets in Medieval and Early Modern Culture* (Princeton University Press, 1994); Nicolas Weill-Parot, "Encadrement ou

dévoilement: L'occulte et le secret dans la nature chez Albert le Grand et Roger Bacon," *Micrologus: Natura, Scienze e Società Medievali* 14 (2006): 151–70; Chiara Crisciani, "Ruggero Bacone e l' 'Aristotele' del *Secretum secretorum*," in *Christian Readings of Aristotle from the Middle Ages to the Renaissance*, ed. Luca Bianchi (Brepols, 2011), 37–64.

24. Hackett, "Roger Bacon on *Scientia Experimentalis*"; Jeremiah Hackett, "Roger Bacon's Concept of Experience: A New Beginning in Medieval Philosophy?," *Modern Schoolman* 86, no. 1/2 (November 2008): 123–46; Hackett, "*Ego Expertus Sum*"; Jeremiah Hackett, "Roger Bacon's *Scientia Experimentalis*: A Review of Bacon's Diverse Treatises," in *The Sciences of Roger Bacon's* Opus maius, ed. Meagan S. Allen (Routledge, 2025), 12–42; Paola Bernardini and Anna Rodolfi, eds., *Roger Bacon's Communia Naturalium: A 13th Century Philosopher's Workshop*, Micrologus 64 (SISMEL Edizioni del Galluzzo, 2014); Nicola Polloni and Yael Kedar, eds., *The Philosophy and Science of Roger Bacon: Studies in Honour of Jeremiah Hackett* (Routledge, 2021).

25. See Power, *Roger Bacon and the Defence of Christendom*; Power, "A Mirror for Every Age"; Power, "Seeking Remedies for Great Danger"; Amanda Power, "Going Among the Infidels: The Medicant Orders and Louis IX's First Mediterranean Campaign," *Mediterranean Historical Review* 25, no. 2 (2010): 187–202; Amanda Power, "Franciscan Advice to the Papacy in the Middle Ages," *History Compass* 5, no. 5 (2007): 1550–75.

26. Joan Cadden, *The Meanings of Sex Difference in the Middle Ages: Medicine, Science, and Culture* (Cambridge University Press, 1993); Joan Cadden, *Nothing Natural Is Shameful: Sodomy and Science in Late Medieval Europe* (University of Pennsylvania Press, 2013); Lorraine Daston and Katharine Park, *Wonders and the Order of Nature, 1150–1750* (Zone, 1998); Pamela O. Long, *Openness, Secrecy, Authorship: Technical Arts and the Culture of Knowledge from Antiquity to the Renaissance* (Johns Hopkins University Press, 2001); Pamela O. Long, *Technology in Mediterranean and European Lands, 600-1600* (Johns Hopkins University Press, 2025).

27. Cecilia Trifogli, *Oxford Physics in the Thirteenth Century (ca. 1250–1270), Motion, Infinity, Place and Time*, Studien und Texte zur Geistesgeschichte des Mittelalters (Brill, 2000); A. George Molland, "Roger Bacon's Appropriation of Past Mathematics," in *Tradition, Transmission, Transformation: Proceedings of Two Conferences on Premodern Science Held at the University of Oklahoma*, ed. F. Jamil Ragep, Sally P. Ragep, and Steven J Livesey (Brill, 1996), 347–65; Simon Oliver, "Robert Grosseteste on Light, Truth and *Experimentum*," *Vivarium* 42, no. 2 (2004): 151–80.

28. Abdelhamid I. Sabra, "The Appropriation and Subsequent Naturalization of Greek Science in Medieval Islam: A Preliminary Statement," *History of Science* 25, no. 3 (1987): 223–43; George Saliba, *Islamic Science and the Making of the European Renaissance* (MIT Press, 2007); F. Jamil Ragep, "Islamic Culture and the Natural Sciences," in *Cambridge History of Science*, vol. 2, *Medieval Science*, ed. David C. Lindberg and Michael H. Shank (Cambridge University Press, 2013), 27–61.

29. E. R. Truitt, "Knowledge and Power: Courtly Science and Political Utility in the Work of Roger Bacon," *Revista Española de Filosofía Medieval* 28, no. 1 (2021): 99–123; H. Darrel Rutkin, "An Idealized Astrological Courtier at a 13th Century Papal and Royal Court: The Case of Roger Bacon," in *De Frédéric II à Rodolphe II: Astrologie, divination et magie dans le cours (XIIIe-XVIIe siècle)*, ed. Jean-Patrice Boudet, Martine Ostorero, and Agostino Paravicini Bagliani, Micrologus 85 (SISMEL Edizioni del Galluzzo, 2017), 53–69.

30. Because much of this book is oriented around the composition, copying, and circulation of texts, I use linguistic descriptors related to the language of textual production:

Latin and Arabic. On the use of language as a descriptor, see Peter Adamson and Richard Taylor, "Introduction," in *Cambridge Companion to Arabic Philosophy*, ed. Peter Adamson and Richard Taylor (Cambridge University Press, 2005), 3–4; Cynthia Robinson, *In Praise of Song: The Making of Courtly Culture in Al-Andalus and Provence, 1005–1134 A.D.* (Brill, 2002), 17; María Rosa Menocal, *The Arabic Role in Medieval Literary History: A Forgotten Heritage* (University of Pennsylvania Press, 1990). When referring broadly to the areas under Muslim or Christian political control, I have chosen to use (Latin) "Christendom" and "Islamdom" for a sense of parity, and because both terms encompass political and cultural realities. "Islamdom" was proposed by Marshall Hodgson to discuss and characterize society, as opposed to religion or culture, but has not been widely taken up. Hodgson also coined "Islamicate," to denote culture, rather than society or religion; this is more widely used, though also widely critiqued. See Marshall G. S. Hodgson, *The Venture of Islam: Conscience and History in a World Civilization*, 3 vols. (University of Chicago Press, 1974); Shahab Ahmed, *What Is Islam? The Importance of Being Islamic* (Princeton University Press, 2015), especially 157–75.

31. Whewell was faculty and eventual head of Trinity College, Cambridge; White was college faculty and president of Cornell University; Kuhn was at Harvard and University of California, Berkeley. Other important actors in this study were situated within institutions like Columbia University and Oxford University. Abigail Boggs and Nick Mitchell, "Critical University Studies and the Crisis Consensus," *Feminist Studies* 44, no. 2 (2018): 432–63; Abigail Boggs, Eli Meyerhoff, Nick Mitchell, and Zach Schwartz-Weinstein, "Abolitionist University Studies: An Invitation," *Abolition University* (blog), accessed August 25, 2024, https://abolition.university/wp-content/uploads/2019/08/Abolitionist-University-Studies_-An-Invitation-Release-1-version.pdf.

32. Whewell, *Architectural Notes on German Churches, with Remarks on the Origin of Gothic Architecture*, 2nd ed. (J. & J. J. Deighton, 1835 [1830]). This revised edition contains additional material on churches in northern France.

33. Besides Kuhn, *The Structure of Scientific Revolutions*, see Duhem, *Le système du monde*; Alexandre Koyré, *From the Closed World to the Infinite Universe* (Johns Hopkins University Press, 1957); Owen Gingerich, *The Eye of Heaven: Ptolemy, Copernicus, Kepler* (American Institute of Physics, 1993); David Wootton, *Galileo: Watcher of the Skies* (Yale University Press, 2010); Robert S. Westman, *The Copernican Question: Prognostication, Skepticism, and Celestial Order* (University of California Press, 2011).

34. To take only a few examples from the past century, Alexandre Koyré, *Études Galiléennes*, 3 vols. (Hermann, 1939); Alexandre Koyré, "Galileo and Plato," *Journal of the History of Ideas* 4 (1943): 400–428; Alistair Crombie, "The Significance of Medieval Discussions of Scientific Method for the Scientific Revolution," in *Critical Problems in the History of Science*, ed. Marshall Clagett (University of Wisconsin Press, 1959), 79–101; John E. Murdoch, "Philosophy and the Enterprise of Science in the Later Middle Ages," in *The Interaction between Science and Philosophy*, ed. Yehuda Elkana and Samuel Sambursky (Humanities Press, 1974), 51–74; Steven Shapin, *The Scientific Revolution* (University of Chicago Press, 1996); Peter Dear, *Revolutionizing the Sciences: European Knowledge and Its Ambitions, 1500–1700* (Princeton University Press, 2001); H. Floris Cohen, *How Modern Science Came into the World: Four Civilizations, One 17th-Century Breakthrough* (Amsterdam University Press, 2010); David Wootton, *The Invention of Science: A New History of the Scientific Revolution* (HarperCollins, 2015).

35. For example, Steven Shapin and Simon Schaffer, *Leviathan and the Air-Pump: Hobbes, Boyle, and the Experimental Life* (Princeton University Press, 1985); Steven Shapin, "The House of Experiment in Seventeenth-Century England," *Isis* 79 (1988): 373–404; Simon

Schaffer, "Glass Works: Newton's Prisms and the Uses of Experiment," in *The Uses of Experiment: Studies in the Natural Sciences*, ed. David Gooding, Trevor Pinch, Simon Schaffer (Cambridge University Press, 1989), 67–104; Pamela H. Smith, *The Body of the Artisan: Art and Experience in the Scientific Revolution* (University of Chicago, 2004); Pamela O. Long, *Artisan/Practitioners and the Rise of the New Sciences, 1400–1600* (University of Oregon, 2011); Mordechai Feingold, "'Experimental Philosophy': Invention and Rebirth of a Seventeenth-Century Concept," *Early Science and Medicine* 21 (2016): 1–28. For a critique of this chronology, see E. R. Truitt, "Making English Scientific: Chaucer, Translation, and the Astrolabe," *Isis* 115, no. 4 (2024): 757–75.

36. Chakrabarty, *Provincializing Europe*; Projit Bihari Mukharji, *Doctoring Traditions: Ayurveda, Small Technologies, and Braided Sciences* (University of Chicago Press, 2016); Pablo F. Gómez, *The Experiential Caribbean: Creating Knowledge and Healing in the Early Modern Atlantic* (University of North Carolina Press, 2017); Julie Livingston, *Self-Devouring Growth: A Planetary Parable as Told from Southern Africa* (Duke University Press, 2019); Suman Seth, *Difference and Disease: Medicine, Race, and the Eighteenth-Century British Empire* (Cambridge University Press, 2018).

CHAPTER 1

Note to epigraph. "Oportet ergo omnia certificari per viam experientiae." Roger Bacon, *Opus majus*, ed. John Henry Bridges, 3 vols. (Clarendon Press, 1897–1900), 6.1, 2:168.

1. "Et per tuas nobis declares litteras quae tibi videntur adhibenda remedia circa illa, quae nuper occasione tanti discriminis intimasti: et hoc quanto secretius poteris facias indilate." Roger Bacon, "Epistola Clementis Papae IV. ad Rogerium Baconem," in *Fr. Rogeri Bacon opera quædam hactenus inedita*, ed. J. S. Brewer (Longman and Green, 1859), 1.

2. Justine Firnhaber-Baker, *House of Lilies: The Dynasty That Made Medieval France* (Basic Books, 2024), 181–206.

3. Christopher I. Beckwith, *Empires of the Silk Road: A History of Central Eurasia from the Bronze Age to the Present* (Princeton University Press, 2009), 184–90.

4. "Et proculdubio Tartari ruperunt portas et transiverunt claustra Alexandri et mundum vastaverunt, unde timendum est multum de eis." Roger Bacon, *Opus tertium*, ed. and trans. Nikolaus Egel (Felix Meiner, 2020), 2.78, 684. Bacon also discussed the Mongols at length in Book 4 of *Opus maius*. See also Davide Bigalli, *I Tartari e l'Apocalisse: Ricerche sull'escatologia in Adamo Marsh e Ruggero Bacone* (La Nuova Italia, 1971), 137–40.

5. Bacon, *Opus tertium* 2.78, ed. Egel, 684. By Bacon's time, the legend that Gog and Magog had been captured and imprisoned by Alexander, which dated back at least to the first century, had become part of the lore around the Antichrist. Zachary A. Matus, *Franciscans and the Elixir of Life: Religion and Science in the Later Middle Ages* (University of Pennsylvania Press, 2017), 79.

6. Amanda Power, *Roger Bacon and the Defence of Christendom* (Cambridge University Press, 2011), 57–59. On Bacon's movement between Paris and Oxford in this period, see Thomas S. Maloney, "Introduction," in Roger Bacon, *Compendium of the Study of Theology*, ed. and trans. Thomas S. Maloney (Brill, 1988), 4–9.

7. Firnhaber-Baker, *House of Lilies*, 202–3.

8. William Chester Jordan, *The French Monarchy and the Jews: From Philip Augustus to the Last Capetians* (University of Pennsylvania Press, 1989); Yossef Schwartz, "Authority,

Control, and Conflict in Thirteenth-Century Paris: Contextualizing the Talmud Trial," in *Jews and Christians in Thirteenth-Century France*, ed. Elisheva Baumgarten and Judah D. Galinsky (Palgrave Macmillan, 2015), 93–110.

9. David Carpenter, *Henry III: The Rise to Power and Personal Rule, 1207–1258* (Yale University Press, 2020); J. R. Maddicott, *Simon de Montfort* (Cambridge University Press, 1994).

10. Bacon, *Opus majus*, 4.4, ed. Bridges, 1:385–86.

11. Power, *Roger Bacon and the Defence of Christendom*, 63.

12. "Et creditur ab omnibus sapientibus quod non sumus multum remoti a temporibus Antichristi, sicut in capitulo de sectis per astronomiam in uno revolutis patet." Bacon, *Opus majus*, 4.4, ed. Bridges, 1:407.

13. Matus, *Franciscans and the Elixir of Life*, 77.

14. Bacon, *Opus majus*, 6.2, ed. Bridges, 2:172.

15. "Et hoc ultimo confirmari potest per hoc quod eisdem personis data est philosophiae plenitudo, quibus et lex Dei, scilicet sanctis patriarchis et prophetis a mundi principio." Bacon, *Opus majus*, 2.9, ed. Bridges, 3:53–54. See Camille Bérubé, *De la philosophie à la sagesse chez Saint Bonaventure et Roger Bacon* (Istituto Storico dei Cappucini, 1976), 69–75.

16. Nikolaus Egel, "Roger Bacon's New Founding of the Sciences in the Thirteenth Century," in *The Philosophy and Science of Roger Bacon: Studies in Honour of Jeremiah Hackett*, ed. Nicola Polloni and Yael Kedar (Routledge, 2021), 125–140, 133.

17. These two types of *sapientia* are related in Bacon's work. Paradoxically, *sapientia* in the first sense—*sapientia perfecta* as it was called by the Franciscans—was possible only after renouncing all learning and instead focusing on contemplation. On *sapientia* and its importance to Bacon and to the Franciscan Order, see Amanda Power, "Seeking Remedies for Great Danger: Contemporary Appraisals of Roger Bacon's Expertise," in *Knowledge, Discipline, and Power in the Middle Ages: Essays in Honour of David Luscombe*, ed. Joseph Canning, Edmund J. King, and Martial Staub (Brill, 2011), 63–78; Power, *Roger Bacon and the Defence of Christendom*, 151–59.

18. Timothy J. Johnson, "'Wisdom Has Built Her House; She Has Set Up Her Seven Pillars': Roger Bacon, Franciscan Wisdom, and Conversion to the Sciences," in *The English Province of the Franciscans (1224–c. 1350)*, ed. Michael J. P. Robson (Brill, 2017), 294–315.

19. "Cum vero jam monstravi proprietatem hujus scientiae secundem se, jam cuilibet patet quod utilissima est haec scientia post morale, et primo ipsi theologiae absolutae propter sensum literalem et spiritualem, in quibus ipsa consistit." Bacon, *Opus majus* 6.12, ed. Bridges, 2:219–20. See also Hackett, "Roger Bacon on *Scientia Experimentalis*," in *Roger Bacon and the Sciences: Commemorative Essays*, ed. Jeremiah Hackett (Brill, 1995), 277–316, 309–11; Power, *Roger Bacon and the Defence of Christendom*, 152–54.

20. Egel, "Roger Bacon's New Founding of the Sciences," 129. See also Jeremiah Hackett, "Moral Philosophy and Rhetoric in Roger Bacon," *Philosophy and Rhetoric* 20, no. 1 (1987): 18–40.

21. "Omnia vero opera artis et naturae ordinantur ad haec opera moralia, et sunt propter ea; nec habent utilitatem nisi secundum quod operibus morum deserviunt. . . . Et haec scientia est domina partium omnium philosohpiae, et utitur eis et imperat, propter utilitates civitatum et regnorum." Bacon, *Opus tertium*, 1.14, ed. Egel, 98.

22. "Quatuor vero sunt maxima comprehendendae veritatis offendicula, quae omnem quemcumque sapientem impediunt, et vix aliquem permittunt ad verum titulum sapientiae pervenire, videlicet fragilis et indignae auctoritatis exemplum, consuetudinis diuturnitas,

vulgi sensus imperiti, et propriae ignorantiae occultatio cum ostentatione sapientiae apparentis. His omnis homo involvitur, omnis status occupatur. Nam quilibet in singulis artibus vitae et studii et omnis negotii tribus pessimis ad eandem conclusionem utitur argumentis, cilicet hoc exemplificatum est per majores, hoc consuetum est, hoc vulgatum est; ergo tenedum." Bacon, *Opus majus*, 1.1, ed. Bridges, 3:2.

23. Egel, "Roger Bacon's New Founding of the Sciences," 125. On Bacon's rejection of the *Sentences* for understanding Scripture, see Beryl Smalley, *The Study of the Bible in the Middle Ages*, 3rd ed. (Blackwell, 1983 [1952]), 329–30. Smalley characterizes Bacon as a "rebellious reactionary, or a reactionary rebel."

24. Power has definitively debunked the association between Bacon and Ilchester, Somerset, for which there is no evidence; see Power, *Roger Bacon and the Defence of Christendom*, 32.

25. "Multam laboravi in scientiis et linguis, et posui jam quadraginta annos postquam didici primo alphabetum." Bacon, *Opus tertium*, 1.20, ed. Egel, 132. For an introduction to the different positions on his birth year, and the consequent implications, see Jeremiah Hackett, "Roger Bacon: His Life, Career, and Works," in *Roger Bacon and the Sciences*, ed. Hackett, 9–23, especially 9–11; David C. Lindberg, *Roger Bacon and the Origins of Perspectiva in the Middle Ages: A Critical Edition and English Translation of Bacon's Perspectiva with Introduction and Notes* (Oxford University Press, 1996), xvii–xviii.

26. On the use of classical, pagan authors' works as introductory teaching texts in this period, see José Antonio Fránquiz, "The Place of Seneca in the Curriculum of the Middle Ages," in *Arts libéraux et philosophie au Moyen Âge* (Librairie J. Vrin, 1969), 1065–72. On medieval education in England, see Nicholas Orme, *Medieval Schools: From Roman Britain to Renaissance England* (Yale University Press, 2006).

27. On Bacon's time in Oxford and Paris, see also "Roger Bacon," in *Dictionary of Scientific Biography* (Charles Scribner's Sons, 2008), 377–85.

28. Part of the reason is that Oxford was not an episcopal seat, unlike Paris, but formed part of the very large bishopric of Lincoln. Olaf Pedersen, *The First Universities: Studium Generale and the Origins of University Education in Europe*, trans. Richard North (Cambridge University Press, 1997), 182–83; Jacques Verger, "The First French Universities and the Institutionalization of Learning: Faculties, Curricula, Degrees," in *Learning Institutionalized: Teaching in the Medieval University*, ed. John Van Engen (University of Notre Dame Press, 2000), 5–20.

29. "[Audivi] diligenter plures, et legi plus quam alius." Roger Bacon, *Compendium Studium Philosophiae*, in *Fr. Rogeri Bacon opera*, ed. Brewer, 394–519, 468.

30. Pedersen, *The First Universities*, 275–76.

31. Roger Bacon, *Compendium of the Study of Philosophy*, ed. and trans. Thomas S. Maloney (Oxford University Press, 2018), 4.87, 84–86. Edmond Nolan and Samuel Abraham Hirsch, eds., *The Greek Grammar of Roger Bacon and a Fragment of His Hebrew Grammar* (Cambridge University Press, 1902); Charles Burnett, *The Introduction of Arabic Learning into England* (The British Library, 1997), 77–79; Power, *Roger Bacon and the Defence of Christendom*, 35, 46; Pam Manix, "Oxford: Mapping the Medieval Jewry," in *The Jews of Europe in the Middle Ages (Tenth to Fifteenth Centuries)*, ed. Christoph Cluse (Brepols, 2004), 405–20. On Bacon's later work with students in Hebrew, see Étienne Anheim, Benoît Grévin, and Martin Morard, "Exégèse judéo-chrétienne, magie, et linquistique: Un recueil de *Notes* inédites attribuées à Roger Bacon," *Archives d'Histoire Doctrinale et Littéraire du Moyen Âge* 68 (2001): 95–154.

32. Robert Steele, "Roger Bacon as Professor: A Student's Notes," *Isis* 20, no. 1 (1933): 53–71.

33. Although Gregory IX proclaimed in 1233 the *ius docendi ubique*—a license that granted the bearer to teach anywhere, and that was invented for the new university of Toulouse—he later suggested that it would not supersede the rights and privileges of the university of Paris to grant its own licenses for teaching. See Steven J. Williams, "Roger Bacon in Context: Empiricism in the High Middle Ages," in *Expertus sum: L'expérience par les sens dans la phlosophie naturelle médiévale*, ed. Thomas Bénatouïl et Isabelle Draelants (SISMEL Edizioni del Galluzzo, 2011), 123–44, 126n6.

34. Verger, "The First French Universities and the Institutionalization of Learning," 5–20; John W. Baldwin, *The Scholastic Culture of the Middle Ages, 1000–1300* (Wayland, 1997 [Heath, 1971]).

35. Jeremiah Hackett, "Roger Bacon, Aristotle, and the Parisian Condemnations of 1270, 1277," *Vivarium* 35 (1997): 283–314.

36. Pedersen, *The First Universities*, 278–79.

37. These lectures are printed in Bacon, *Opera hactenus inedita Rogeri Baconi*, ed. Robert Steele (Clarendon Press, 1920), fascs. 7, 8, 10–15. See also Steele, "Roger Bacon as professor"; Irène Rosier-Catach, "Roger Bacon and Grammar," in *Roger Bacon and the Sciences*, ed. Hackett, 67–102; Alain de Libera, "Roger Bacon et la logique," in *Roger Bacon and the Sciences*, ed. Hackett, 103–32; Rega Wood, "Roger Bacon: Richard Rufus' Successor as a Parisian Physics Professor," *Vivarium* 35, no. 2 (1997): 222–50; Power, *Roger Bacon and the Defence of Christendom*, 37.

38. The exact date of Bacon's entrance to the Order is unknown, but due to a comment in *Opus tertium* implying that Bacon had been a Franciscan for a decade prior to Gui's elevation to the papacy the mid-1250s is used; see Power, *Roger Bacon and the Defence of Christendom*, 52n103.

39. It is also possible that the two men met earlier, in 1257; see Norbert Kamp, *Enciclopedia dei Papi*, 3 vols. (Istituto della Enciclopedia Italiana, 2000), 2:401–11, 409.

40. Lynn Thorndike Jr., *A History of Magic and Experimental Science*, 8 vols. (Columbia University Press, 1923), 2:622–23; Power, *Roger Bacon and the Defence of Christendom*, 63–64.

41. "Tuae devotionis litteras gratanter recepimus: sed et verba notavimus diligenter quae ad explanationem earum dilectus filius G. dictus Bonecor, Miles, viva voce nobis proposuit, tam fideliter quam prudenter." Roger Bacon, "Epistola Clementis Papæ IV. ad Rogerium Baconem," in *Fr. Rogeri Bacon opera*, ed. Brewer, 1.

42. Bonecor had been sent to the curia in March 1266 and was still there in July when Clement wrote to Bacon; *Calendar of the Patent Rolls Preserved in the Public Record Office, Henry III (1258–1266)* (Public Record Office, 1910), 566–67, 633. See Power, *Roger Bacon and the Defence of Christendom*, 63n145, and 68 on this point.

43. "Scriptum de bona littera nobis mittere . . . et hoc quanto secretius poteris facias." Bacon, "Epistola Clementis Papæ IV. ad Rogerium Baconem," 1. See also Power, *Roger Bacon and the Defence of Christendom*, 67–69 for a more detailed account of the correspondence between Bacon and Gui de Foulques.

44. "Misi igitur fratri meo diviti in terra mea, qui ex parte regis consistens, cum matre mea, et fratribus, et tota familia, exulavit, et pluries hostibus deprehensus se redemis pecunia; et ideo etiam usque ad hunc diem habui responsum ab eo." Bacon, *Opus tertium*, 1.3, ed. Egel, 32.

45. On the Franciscan prohibition against disseminating work outside the Order without permission, see Michael Bihl, ed., "Statuta generalia ordinis edita in capitulis generalibus celebratis Narbonae an. 1260, Assisii an. 1279 atque Parisiis an. 1292," *Archivum Franciscanum Historicum*, 34, no. 1–2 (1941): 13–91, 73; Power, *Roger Bacon and the Defence of Christendom*, 58.

46. "Nam, teste deo, quando vestre celesitudini obtuli me paratum, intellexi pro scripturis faciendis, non tunc factis." Roger Bacon, "Epistola Fratris Rogeri Baconi," in *Lettera a Clemente IV*, ed. and trans. P. Efrem Bettoni (Biblioteca francescana provinciale, 1964), 82.

47. Thorndike, *History of Magic*, 2:625. He later writes, of Bacon, "For a man so easily discouraged one cannot feel much sympathy." Thorndike, *History of Magic*, 2:626. For context, Thorndike's own eight-volume *History of Magic and Experimental Science* took him decades to complete and runs to close to ten thousand pages.

48. Egel, "Roger Bacon's New Founding of the Sciences," 125.

49. Thorndike, *History of Magic*, 2:622–24. Bacon also likely had written an earlier version of what became the fifth book of the *Great Work*, on *perspectiva* or optics, and revised that material for Clement IV. See Jeremiah Hackett, "Roger Bacon's Various Texts on *Perspectiva* in His Later Works: The Significance of Two New Versions of His *Perspectiva*," in *Arabic and Latin Theory of Perspective*, ed. Agostino Paravicini Bagliani, Micrologus 29 (SISMEL Edizioni del Galluzzo, 2021), 81–104. The chronology of Bacon's works has been a matter of discussion. For example, see David Lindberg, "Introduction," to *Roger Bacon's Philosophy of Nature: A Critical Edition, with English Translation, Introduction, and Notes, of De Multiplicatione Specierum and De Speculis Comburentibus*, ed. and trans. David C. Lindberg (Clarendon Press, 1983), xxiv–xxvi; Eugenio Massa, *Ruggero Bacone etica e poetica nella storia dell' "Opus maius"* (Edizioni di Storia e Letteratura, 1955), 7–23; Franco Alessio, *Mito e scienza in Ruggero Bacone* (Ceschina, 1957), 295–315; Stewart Easton, *Roger Bacon and His Search for a Universal Science: A Reconsideration of the Life and Work of Roger Bacon in the Light of His Own Stated Purposes* (Columbia University Press, 1952), 157–66.

50. Bacon, "Epistola Fratris Rogeri Baconi," 80–94, cited in Power, *Roger Bacon and the Defence of Christendom*, 70n176. Power also notes the rather rudimentary diagrams in early copies of Bacon's work, suggesting that he had not received the necessary funds to hire a scribe to draw them correctly.

51. Power, *Roger Bacon and the Defence of Christendom*, 72.

52. "Et in hac tabula est mirum artificium astronomiae, et summa legis intelligendiae utilitas, et omnium festorum legalium, quam qui nescit numquam potest scire intellectum legis, ut oportet, nec cum Judaeis conferre de talibus, nec eis persuadere utiliter." Bacon, *Opus tertium*, 1.56, ed. Egel, 460. See also C. Philipp E. Nothaft, "Duking It Out in the Arena of Time: Chronology and the Christian-Jewish Encounter (1100–1600)," *Medieval Encounters* 22 (2016): 213–35, 224–25.

53. "Sollicitavi multos et magnos; et aliquorum faciem bene cognoscitis, sed not mentem; et dixi quod negotium quoddam vestrum debuit tractari in Francia per me, licet ilud non expressi, cujus executio indiget pecunia magna. . . . Etiam mihi non credebant amici, quia non potui eis negotium explicare." Bacon, *Opus tertium*, 1.3, ed. Egel, 32. On these excuses see Thorndike, *History of Magic*, 2:623.

54. Bonaventure, master general of the Franciscan Order, was hostile to both alchemy and astral science. See "Roger Bacon," in *Dictionary of Scientific Biography*, 378.

55. A. G. Little, "Roger Bacon's Works with References to the MSS, and Printed Editions," in *Roger Bacon: Essays*, ed. A. G. Little (Clarendon Press, 1914), 375–425, 393. The

position of astrology as a subject of study in the university was a matter of debate. See Jeremiah Hackett, "Aristotle, *Astrologia*, and Controversy at the University of Paris (1266–1274)," in *Learning Institutionalized*, ed. Van Engen, 69–110.

56. Power, "Seeking Remedies for Great Danger," 75.

57. Power, *Roger Bacon and the Defence of Christendom*, 72. Matus states that Bacon neglected to show his work to his superiors beforehand and that this is why a fourteenth-century chronicle reports that he was censured by his Order; see Matus, *Franciscans and the Elixir of Life*, 5. However, the allegation of Bacon's censure has been largely overblown and it is not even clear that it is an accurate rendering of events. See Power, "Seeking Remedies for Great Danger."

58. I discuss the impact of Bacon's ideas in Chapter 3. However, for evidence that his work reached the curia and was read by other scholars, see David C. Lindberg, "Lines of Influence in Thirteenth Century Optics: Bacon, Witelo, and Pecham," *Speculum* 46, no. 1 (1971): 66–83, 73–75; Agostino Paravicini Bagliani, *The Pope's Body*, trans. David S. Peterson (University of Chicago Press, 2000), xiii–xv, 196–98; Paravicini Bagliani, *Medicina e scienze della natura alla corte dei papi nel Duocento* (Centro Italiano di Studi Sull'alto Medioevo, 1991), 117–75.

59. "Haec ergo sola novit perfecte experiri quid potest fieri per naturam, quid per artis indstriam, quid per fraudem, quid volunt et somniant carmina conjurationes invocationes deprecationes sacrificia, quae sunt magica, et quid fit in illis ut tollatur omnis falsitats et sola veritas artis teneatur. Haec sola docet considerare omnes insanias magicorum, non ut confirmentur sed ut vitentur, sicut logica considerat sophisticam artem." Bacon, *Opus majus*, 6.2, ed. Bridges, 2:172.

60. "Et quia Scientia Experimentalis a vulgo studentium est penitus ignorata." Bacon, *Opus majus*, 6.2, ed. Bridges, 2:172.

61. The definition of magic in the thirteenth century was contested and contingent. For an overview, see Richard Kieckhefer, *Magic in the Middle Ages*, 3rd ed. (Cambridge University Press, 2021). Nevertheless, Bacon was opposed to magic and conceived *scientia experimentalis* as doctrinally and intellectually legitimate. Nicolas Weill-Parot, "Knowability and the Power of *Ars*: Roger Bacon's Utopian Science as Anti-Magic," in *The Sciences of Roger Bacon's* Opus Maius, ed. Meagan S. Allen (Routledge, 2025), 130–46.

62. Roger Bacon, *Epistola de secretis*, in *Fr. Rogeri Bacon opera*, ed. Brewer, 523–51.

63. Jeremiah Hackett, "Roger Bacon's Concept of Experience: A New Beginning in Medieval Philosophy?," *Modern Schoolman* 86, no. 1/2 (November 2008): 123–46, 126–27.

64. See Isidore of Seville on testing balsam for purity, *Etymologiarum*, ed. W. M. Lindsay, 2 vols. (Clarendon Press, 1911), 17.8.14; see also E. R. Truitt, "The Virtues of Balm in Late Medieval Literature," *Early Science and Medicine*, 14, no. 6 (2009): 711–36, esp. 718–24. On later medieval experimentation, see Michael McVaugh, "Determining a Drug's Properties: Medieval Experimental Protocols," *Bulletin of the History of Medicine* 91, no. 2 (2017): 183–209.

65. Bacon, *Opus majus* 6.1, ed. Brewer, 2:169–70. This is an echo of the Augustinian doctrine of illumination, as well as al-Kindī's assertion that the search for scientific truth and morality were connected. See Peter Adamson, *Al-Kindī* (Oxford University Press, 2006); Thérèse-Anne Druart, "Al-Kindī's Ethics," *Review of Metaphysics* 47, no. 2 (1993): 329–57; Hackett, "Roger Bacon on *Scientia Experimentalis*," 284–85.

66. "Duo enim sunt modi cognoscendi, scilicet per argumentum et experimentum. Argumentum concludit et facit nos concedere conclusionem, sed non certificat neque

removet dubitationem ut quiescat animus in intuit veritatis, nisi eam inveniat via experientiae; quia multi habent argumenta ad scibilia, sed quia non habent experientiam, negligent ea, nec vitant nociva nec persequuntur bona. Si enim aliquis homo qui nunquam vidit ignem probavit per argumenta sufficientia quod ignis comburit et laedit res et destruit, nunquam propter hoc quiesceret animus audientis, nec ignem vitaret antequam poneret manum vel rem combustibilem ad ignem, ut per experientiam probaret quod argumentum edocebat. Sed assumpta experientia combustionis certificatur animus et quiescit in fulgore veritatis. Ergo argumentum non sufficit, sed experientia." Bacon, *Opus majus*, 6.1, ed. Bridges, 2:167–68.

67. "Quod ergo dicit Aristoteles quod demonstratio syllogismus est faciens scire, intelligendum est si experientia comitetur, et non de nuda demonstratione. Quod etiam dicit primo Metaphysicae, quod habentes rationem et causam sunt sapientiores expertis, loquitur de expertis qui solum noscunt nudam veritatem sine causa. Sed hic loquor de experto, qui rationem et causam novit per experientiam. Et hi sunt perfecti in sapientia." Bacon, *Opus majus*, 6.1, ed. Bridges, 2:168.

68. On Aristotle's distinction, see *Metaphysica*, 981a23–981b9, in *Metaphysics*, vol. 1, *Books 1–9*, trans. Hugh Tredennick, Loeb Classical Library 271 (Harvard University Press, 1933), 4–7.

69. Bacon, *Opus majus*, 6.2, ed. Bridges, 2:172–74, although he continues discussing the rainbow over the next ten chapters as well. Bacon's discussion of the rainbow has invited significant scholarly examination; in particular, see David C. Lindberg, "Roger Bacon's Theory of the Rainbow: Progress or Regress?," *Isis* 57, no. 2 (1966): 235–49; Lindberg, "Lines of Influence"; Hackett, "Roger Bacon on *Scientia Experimentalis*."

70. See William R. Newman, "An Overview of Roger Bacon's Alchemy," in *Roger Bacon and the Sciences*, ed. Hackett, 317–36; Faye Getz, "Roger Bacon and Medicine: The Paradox of the Forbidden Fruit and the Secrets of Long Life," in *Roger Bacon and the Sciences*, ed. Hackett, 337–64; Paravicini Bagliani, "Ruggero Bacone, Bonifacio VIII, e la teoria della *prolongatio vitae*," in *Medicina e scienze*, ed. Paravicini Bagliani, 327–62; Meagan S. Allen, *Roger Bacon and the Incorruptible Human, 1220–1292: Alchemy, Pharmacology, and the Desire to Prolong Life*, Palgrave Studies in Medieval and Early Modern Medicine (Palgrave Macmillan, 2023).

71. Bacon, *Opus majus*, 4.4, ed. Bridges, 1:230–31.

72. "Nam ea quae in coelestibus sunt considerantur per instrumenta visualia, ut Ptolemaeus et caeteri docent astronomi. . . . certificantur per modos videndi in instrumentis." Bacon, *Opus majus*, 5.1, ed. Bridges, 2:2.

73. "Experimentator autem perfectus . . . Et tunc thesarum unius regis valeret hoc instrumentum et cessarent instrumenta astronomiae, et horologia, et esset pulcherrimum spectaculum sapientiae." Bacon, *Opus majus*, 6.12, ed. Bridges, 2:203. Bacon's words may have inspired Richard of Wallingford, abbot of St. Alban's Abbey in England, just several decades later. The son of a blacksmith, he combined his knowledge of materials and hands-on knowledge with advanced mathematics to create a complicated mechanical device that modeled celestial and planetary movement, and could be used to calculate calendar dates and tides. See John D. North, *God's Clockmaker: Richard of Wallingford and the Invention of Time* (Hambledon, 2005); John D. North, "Monasticism and the First Mechanical Clocks," in *The Study of Time II: Proceedings of the Second Conference of the International Society for the Study of Time, Lake Yamanaka, Japan*, ed. J. T. Fraser and Nathaniel Lawrence (Springer Verlag, 1975): 381–98.

74. Allen, *Roger Bacon and the Incorruptible Human*, 4.

75. Matus, *Franciscans and the Elixir of Life*, 8–9.

76. Bacon, *Opus tertium*, ch. 22, ed. Egel, 78–86; Matus, *Franciscans and the Elixir of Life*, 43.

77. "Sed experimentator fidelis consideravit aquilam, et cervum, et serpentem, et phoenicem et multa prolongare vitam et renovare juventutem, et scit quod haec dantur brutis ad instructionem hominis; propter quod excogitavit vias nobiles ad hoc, et ussit alkimiae practicae praeparare ei corpus aequalist complexionis, ut hic experimentator uteretur eo." Bacon, *Opus tertium*, 1.13, ed. Egel, 88.

78. Matus, *Franciscans and the Elixir of Life*, 50.

79. Leah DeVun, *Prophecy, Alchemy, and the End of Time: John of Rupescissa in the Late Middle Ages* (Columbia University Press, 2009), 84–85. Meagan Allen argues that Bacon was not very influential; see *Roger Bacon and the Incorruptible Human*, 213.

80. Power, *Roger Bacon and the Defence of Christendom*, 164–207; Raoul Carton, *L'Expérience physique chez Roger Bacon* (J. Vrin, 1924), 62–71.

81. "Oleum citrunum petroleum, id est, oriens ex petra, comburit quicquid occurrit, si rite praeparetur . . . eam aqua no extinguit." Bacon, *Opus majus* 6.12, ed. Bridges 2:217–18.

82. "Nec posset civitas nec exercitus sustinere." Bacon, *Opus majus* 6.12, ed. Bridges, 2:218.

83. "Et hoc in duobus consistit; scilicet in cognitio futurorum praeteritorum et praesentium, et in operibus admirandis quibus excedit astronomiam judiciariam vulgatam in potestate judicandi . . . Haec autem scientia definitiones et vias adinvenit, per quas expedite ad omnem quaestionem respondeat . . . et per quas ostendat nobis figurationes coelestium virtutum; et impressiones coelestium in hoc mundo, sine difficultate astronomiae vulgatae." Bacon, *Opus majus*, 6.12, ed. Bridges, 2:215–16. See also Jeremiah Hackett, "*Ego Expertus Sum*: Roger Bacon's Science and the Origins of Empiricism," in *Expertus sum: L'expérience par les sens dans la philosophie naturelle médiévale*, ed. Thomas Bénatouïl and Isabelle Draelants, Micrologus 40 (SISMEL Edizioni del Galluzzo, 2011), 145–73.

84."Et contra inimicos reipublicae adinvenerunt magnas artes, ut sine ferror, et absque eo quo tangerent aliquem, destruerent comnes resistentes, et eorum sunt multa genera." Bacon, *Opus majus* 6.12, ed. Bridges, 2:217. Hilary Carey refers to Bacon's weaponized natural knowledge in the harshest terms: see "Astrology and Antichrist in the Later Middle Ages," in *Time and Eternity: The Medieval Discourse*, ed. Gerhard Jaritz and Gerson Moreno-Riaño (Brepols, 2003), 515–35; for a different view, see Patricia Clare Ingham, *The Medieval New: Ambivalence in an Age of Innovation* (University of Pennsylvania Press, 2015), 64.

85. "Sicut Alexander doctrina Aristotelis ut historiae narrant, basilisci speciem venenosam positi super murum civitatis ad interficiendum exercitum per corpora magna polita retorsit in eandem civitatem, ut per proprium destrueretur venenum." Bacon, *Opus majus*, 4.4.7, ed. Bridges, 1:143.

86. "Et hoc deberet ecclesia considerare contra infidels et rebelles, ut parcatur sanguine Christiano, et maxime propter future pericula in temporibus Antichristi, quibus cum Dei gratia facile esset obviare, si praelati et principes stadium promoverent et secreta naturae et artis indagarent." Bacon, *Opus majus*, 6.12, ed. Bridges, 2:222, emphasis mine. On the relationship between Bacon's scientific ideas and belief in the apocalypse, see Zachary Matus, "Reconsidering Roger Bacon's Apocalypticism in Light of His Alchemical and Scientific Thought," *Harvard Theological Review* 105, no. 2 (2012): 189–222.

87. "Sed infideles occupantur his insaniis et confidunt in eis, et crediderunt Christianos uti talibus in operibus miraculorum." Bacon, *Opus majus*, 6.12, ed. Bridges, 2:221.

88. William of Rubruck, *The Journey of William of Rubruck to the Eastern Parts of the World, 1253–55*, ed. and trans. William Woodville Rockhill (Hakluyt Society, 1900), 151–52, 156, 187, 239–40; Bigalli, *I Tartari e l'Apocalisse*; E. R. Truitt, *Medieval Robots: Mechanism, Magic, Nature, and Art* (University of Pennsylvania Press, 2015), 32–34.

89. Power, *Roger Bacon and the Defence of Christendom*, 62.

90. "Quod utilissima est haec scientia post morale." Bacon, *Opus majus*, 6.12, ed. Bridges, 2:221.

CHAPTER 2

Note to epigraph. "Narrabo igitur nunc primo opera artis et naturae miranda, ut posea causas et modum eorum assignem; in quibus nihil magicum est; ut videatur quod omnis magica potestas sit inferior his operibus et indigna." Roger Bacon, *Epistola de secretis operibus artis et naturae, et de nullitate magiae*, in *Fr. Rogeri Bacon opera quædam hactenus inedita*, ed. J. S. Brewer (Longman, Green, 1859), 523–51, 532.

1.Roger Bacon, *Opus majus*, ed. John Henry Bridges, 3 vols. (Clarendon Press, 1897–1900), 4.4.16, 2:253 (Ptolemy); 5.1.1, 2:419 (Aristotle); 6.1, 2:585 (Pliny); see also Jeremiah Hackett, "*Ego Expertus Sum*: Roger Bacon's Science and the Origins of Empiricism," in *Expertus sum: L'expérience par les sens dans la philosophie naturelle médiévale*, ed. Thomas Bénatouïl and Isabelle Draelants, Micrologus 40 (SISMEL Edizioni del Galluzzo, 2011), 145–73; Steven J. Williams, "Roger Bacon in Context: Empiricism in the High Middle Ages," in *Expertus sum*, ed. Bénatouïl and Draelants, 123–44, 131. Some of the material under discussion in this chapter first appeared in the article "Knowledge and Power: Courtly Science and Political Utility in the Work of Roger Bacon," *Revista Española de Filosofía Medieval* 28, no. 1 (2021): 99–123.

2. Roger Bacon, *Secretum secretorum*, in *Opera hactenus inedita Rogeri Baconi*, fasc. 5, *Secretum secretorum cum glossis et notulis: Tractatus brevis et utilis ad declarandum quedam obscure dicta Fratris Rogeri*, ed. Robert Steele (Clarendon Press, 1920), 9–10.

3. Steven Marrone, "Metaphysics and Science in the Thirteenth Century: William of Auvergne, Robert Grosseteste, and Roger Bacon," in *Routledge History of Philosophy*, vol. 3, *Medieval Philosophy*, ed. John Marenbon (Routledge, 1998), 250–72.

4. Robert Grosseteste, *De cometis*, in *Die philosophischen Werke des Robert Grosseteste, Bischofs von Lincoln*, ed. Ludwig Baur (Aschendorff Verlag, 1912); Robert Grosseteste, *Commentarius in Posteriorum analyticorum libros*, ed. Pietro Rossi, Unione Accademica Nazionale (Leo S. Olschki, 1981); Alistair C. Crombie, *Robert Grosseteste and the Origins of Experimental Science, 1100–1700* (Clarendon Press, 1953), 62–74; Jeremiah Hackett, "Roger Bacon on *Scientia Experimentalis*," in *Roger Bacon and the Sciences: Commemorative Essays*, ed. Jeremiah Hackett (Brill, 1997), 277–315, 287; Williams, "Roger Bacon in Context," 127–28; Giles E. M. Gasper, Brian K. Tanner, Sigbjørn Sønnesyn, Nader el-Bizri, "Travelling Optics: Robert Grosseteste and the Optics Behind the Rainbow," in *Medieval Science in the North: Travelling Wisdom, 1000–1500*, ed. Christian Etheridge and Michele Campopiano (Brepols, 2021), 45–76.

5. Jeremiah Hackett, "*Scientia experimentalis*: From Robert Grosseteste to Roger Bacon," in *Robert Grosseteste: New Perspectives on His Thought and Scholarship*, ed. James McEvoy

(Brepols, 1995), 89–119, 107–9; Cecilia Panti, "The Theological Use of Science in Robert Grosseteste and Adam Marsh According to Roger Bacon: The Case Study of the Rainbow," in *Robert Grosseteste and the Pursuit of Religious and Scientific Learning in the Middle Ages*, ed. Jack P. Cunningham and Mark Hocknull (Springer, 2016), 143–63, esp. 145–51; Christina van Dyke, "The Truth, the Whole Truth, and Nothing but the Truth: Robert Grosseteste on Universals (and the *Posterior Analytics*)," *Journal of the History of Philosophy* 48, no. 2 (2010): 153–70.

6. Pieter Buellens, "Robert Grosseteste and the Fluid History of the Latin 'Nichomachean Ethics,'" *Revista Española de Filosofía Medieval* 30 (2023): 177–98; A. C. Dionisotti, "On the Greek Studies of Robert Grosseteste," in *The Uses of Greek and Latin: Historical Essays*, ed. A. C. Dionisotti, Anthony Grafton, and Jill Kraye (The Warburg Institute, University of London Press, 1988), 19–39; Cecilia Panti, "In the Margins of the *Posterior Analytics*, Robert Grosseteste and the 'Latin Philoponus,'" *Revista Española Filosofía Medieval* 30, no. 1 (2023): 15–52.

7. Roger Bacon, *Compendium studii philosophiae*, in *Fr. Rogeri Bacon opera*, ed. Brewer, 394–519, 469: "dominus Robertus, . . . neglexit omnio libros Aristotelis et viae eorum, et per experientiam propriam, et auctores alios, et per alias scientias negotiatus est in sapientialibus Aristotelis; et melius centies milesies scivit et scriptis illa de quibus libri Aristotelis loquuntur, quam in ipsius perversis translationibus capi possunt."

8. "Nam semper posteriores addiderunt ad opera priorum, et multa correxerunt, et plura mutaverunt, sicut maxime per Aristotelem patet, qui omnes sententias praecedentium discussit." Bacon, *Opus majus*, 1.6, ed. Bridges, 1:14.

9. Bacon, *Opus majus*, 1.6, ed. Bridges, 1:16.

10. Charles Burnett, "Michael Scot and the Transmission of Scientific Culture from Toledo to Bologna via the Court of Frederick II Hohenstaufen," in *Le scienze alla corte di Federico II*, Micrologus 2 (SISMEL Edizioni del Galluzzo,1994): 101–26; Piero Morpurgo, "*Philosophia naturalis* at the Court of Frederick II: From the Theological Method to the *ratio secundum physicam* in Michael Scot's *De anima*," in *Intellectual Life at the Court of Frederick II Hohenstaufen*, ed. William Tronzo, Studies in the History of Art 44 (National Gallery of Art, 1994), 241–50; Danielle Jacquart, "Le physiognomie à l'époque de Frédéric II: Le traité de Michel Scot," in *Recherches médiévales sur la nature humaine: Essais sur la réflexion médicale (XIIe–XVe s.)*, Micrologus 63 (SISMEL Edizioni del Galluzzo, 1994): 19–37.

11. Martina Giese, "Die Tierhaltung am Hof Kaiser Friedrichs II, zwischen Tradition und Innovation," in *Herrschaftsraüme, Herrschaftspraxis und Kommunikation zur Zeit Kaiser Friedrichs II*, ed. Knut Göring, Jan Keupp, and Theo Broekman, 2 vols. (Herbert Utz Verlag, 2008), 2:121–72, esp. 124–26. Katharine Park generously shared her chapter draft on Frederick II's falconry with me, and it is from her that I learned of this study on Frederick's menagerie.

12. "In scribendo etiam Aristotilem, ubi oportuit, secuti sumus. In pluribus enim, sicut experientia didicimus, maxime in naturis quarundam avium, discrepare a veritate videtur. Propter hoc non sequimur principem philosophorum in omnibus, raro namque aut nunquam venationes avium exercuit, sed nos semper dileximus et exercuimus. De multie vero, que narrat in libro animalium, dicit quosdam sic dixisse, sed id, quod quidam sic dixerunt, nec ipse forsan vidit, nec dicentes viderunt, fides enim certa non provenit ex auditu." Frederick II, *De arte venandi cum avibus*, ed. Carl Arnold Willemsen (In aedibus insulae, 1942), prologue. On this text, see Martina Giese, "The *De Arte venandi cum avibus* of Emperor Frederick II," in *Raptor and Human: Falconry and Bird Symbolism Through the Millennia on a*

Global Scale, ed. Karl-Heinz Gersmann and Oliver Grimm, 4 vols. (Wachholtz-Murmann, 2018), 1:1459–69.

13. "Utilitas est magna": Frederick II, *De arte venandi avibus*, prologue.

14. Baudoin Van den Abeele, "Medieval Western and Eastern Treatises on Falconry in Comparison," in *Falconry in the Mediterranean Context During the Pre-Modern Era*, ed. Charles Burnett and Baudoin Van den Abeele (Droz, 2021), 313–31; Van den Abeele, *La fauconnerie au Moyen Age: Connaissance, affaitage et médicine des moiseaux de chasse d'après les traités latins* (Klincksieck, 1994).

15. "In partibus autem Indiae . . . et propter hoc magica naturalis potissimum ibi viguit, et viget quam de causam sunt ibi super experimentatores multi, et rerum mirabilium per huiusmodi peritiam effectores." William of Auvergne, *De universo*, 2.3.23, in *Opera omnia*, ed. Peter Aubouin, 2 vols. (Paris, 1674), 1:1065, col. 1. See also Antonella Sannino, "Guillaume d'Auvergne e i *libri experimentorum*," in *Expertus sum*, ed. Bénatouïl and Draelants, 67–88. On India as a source of particular natural marvels and powerful natural objects with occult powers, see Mary Baine Campbell, *Witness and the Other World: Exotic European Travel Writing, 400–1600* (Cornell University Press, 1988); Lorraine Daston and Katharine Park, *Wonders and the Order of Nature, 1150–1750* (Zone, 1998).

16. "Sed aliqui viri . . . videntes quod per textum Aristotelis et commentatorem suum non potuerunt scire naturalem philosophiam, convertunt se ad alias sciencias naturales septem, et ad mathematicas, et ad alios autores naturalis philosophie . . . et sic pervenerunt ad noticiam naturalium, de quibus Aristoteles in libris vulgatis et eius expositor non possunt satisfacere studio naturali." Roger Bacon, *Communia naturalium*, ed. Robert Steele (Clarendon Press, 1909), 2.12–13; see also Williams, "Roger Bacon in Context," 130–32; Cecilia Trifogli, *Oxford Physics in the Thirteenth Century (ca. 1250–1270)*, Studien und Texte zur Geistesgeschichte des Mittelalters 72 (Brill, 2000), 159–64.

17. Isabelle Draelants, "Expérience et autorités dans la philosophie naturelle d'Albert le Grand," in *Expertus sum*, ed. Bénatouïl and Draelants, 89–122. Bacon was not impressed with Albert's scholarship. See Jeremiah Hackett, "The Attitude of Roger Bacon to the *Scientia* of Albertus Magnus," in *Albertus Magnus and the Sciences: Commemorative Essays, 1980*, ed. James A. Weisheipl (Pontifical Institute of Medieval Studies, 1980), 53–72; David J. Collins, S.J., *Disenchanting Albert the Great: The Life and Afterlife of a Medieval Magician* (Pennsylvania State University Press, 2024), 23–45.

18. Williams, "Roger Bacon in Context."

19. See Jeremiah Hackett, "From *Sapientes antiqui* at Lincoln to the New *Sapientes moderni* at Paris c. 1260–1280: Roger Bacon's Two Circles of Scholars," in *Robert Grosseteste and the Pursuit of Religious and Scientific Learning in the Middle Ages*, ed. Jack P. Cunningham and Mark Hocknull, Studies in the History of Philosophy of Mind 18 (Springer, 2016), 120–28; Steven J. Williams, "Roger Bacon and His Edition of the Pseudo-Aristotelian *Secretum secretorum*," *Speculum* 69, no. 1 (1994): 57–73, 59; Charles Burnett, *The Introduction of Arabic Learning into England* (The British Library, 1997).

20. Even Hovdaugen, "'Una et eadem': Some Observations on Roger Bacon's Greek Grammar," in *De ortu grammaticae: Studies in Medieval Grammar and Linguistic Theory in Memory of Jan Pinborg*, ed. Geoffrey L. Bursill-Hall, Sten Ebbesen, and E. F. Konrad Koerner (John Benjamins Pub. Co., 1990), 117–31; Edmond Nolan and Samuel Abraham Hirsch, eds., *The Greek Grammar of Roger Bacon and a Fragment of His Hebrew Grammar* (Cambridge University Press, 1902). On Bacon's work in Hebrew, see Etienne Anheim, Benoît Grévin, and Martin Morard, "Exégèse judéo-chrétienne, magie, et linquistique: Un recueil de *Notes*

inédites attribuées à Roger Bacon," *Archives d'Histoire Doctrinale et Littéraire du Moyen Âge* 68 (2001): 95–154; Horst Weinstock, "Roger Bacon und das 'hebräische Alphabet,'" *Aschkenas* 2 (1992): 15–48.

21. "Et vis huius rei stat in hoc: ut homo sciat legere Graecum et Hebraeum et cetera." Roger Bacon, *Compendium of the Study of Philosophy*, ed. and trans. Thomas S. Maloney (Oxford University Press, 2018), 6.86, 84–85.

22. "Et secundo considerandum est quod interpretes non habuerunt vocabula in Latino pro scientiis transferendis, quia no fuerunt primo compositae in lingua Latina." Roger Bacon, *Opus majus*, 3.1, ed. Bridges, 1:67. Richard Lemay, "Roger Bacon's Attitude Toward the Latin Translations and Translators of the Twelfth and Thirteenth Centuries," in *Roger Bacon and the Sciences*, ed. Hackett, 25–48, esp. 37–39; Burnett, *The Introduction of Arabic Learning into England*, 76–79; Beryl Smalley, *The Study of the Bible in the Middle Ages*, 3rd ed. (Blackwell, 1983 [1952]), 331–35. On Bacon's view of translation, language, and equivocation, see Patricia Clare Ingham, *The Medieval New: Ambivalence in an Age of Innovation* (University of Pennsylvania Press, 2015), 60–62.

23. See Roger Bacon, "Introduction" to *Quaestiones supra De Plantis*, in *Opera hactenus inedita Rogeri Baconi*, ed. Steele, fasc. 11, xviii, xxv–xxiv; Lemay, "Roger Bacon's Attitude Toward the Latin Translations," 37–39; Power, *Roger Bacon and the Defence of Christendom*, 37.

24. Nicola Polloni, "Disentangling Roger Bacon's Criticism of Medieval Translations," in *Early Thirteenth-Century English Franciscan Thought*, ed. Lydia Schumacher (De Gruyter, 2021), 261–82.

25. Power, *Roger Bacon and the Defence of Christendom*, 46; Lemay, "Roger Bacon's Attitude Toward the Latin Translators," 46.

26. *De aspectibus* is extant only in Latin version. *De radiis*, attributed to al-Kindī more recently, was likely a text by a Latin author in the mid-thirteenth century. On the authorship of *De radiis*, see Sylvain Matton, "An Irradiation of Latin Grammarians; or, The *De radiis* Is Not by al-Kindī," *Magic, Ritual, and Witchcraft* 17, no. 3 (2022): 437–55.

27. Peter Adamson, *Al-Kindī* (Oxford University Press, 2006), 167–172; Elaheh Kheirandish, "The Mixed Mathematical Sciences: Optics and Mechanics in the Islamic Middle Ages," in *Cambridge History of Science*, vol. 2, *Medieval Science*, ed. David C. Lindberg and Michael H. Shank (Cambridge University Press, 2013), 84–108, 87.

28. David C. Lindberg, *Theories of Vision from Al-Kindi to Kepler* (University of Chicago Press, 1976), 18–57.

29. Ibn al-Haytham, *Kitāb al-manāẓir*, ed. Abdelhamid I. Sabra (The National Council for Culture, Arts, and Letters of Kuwait, 1983); Crombie, *Robert Grosseteste and the Origins of Experimental Science*, 146–48; David C. Lindberg, "The Western Reception of Arabic Optics," in *The Encyclopedia of the History of Arabic Science*, vol. 3, *Technology, Alchemy, and Life Sciences*, ed. Roshdi Rashed and Régis Morelon, 3 vols. (Routledge, 1996), 2:716–29; Lindberg, *Theories of Vision*, 58–86. Ibn al-Haytham wrote at least fifteen works on optics; see Abdelhamid I. Sabra, *The Optics of Ibn al-Haytham: Books I–III, On Direct Vision*, 2 vols. (The Warburg Institute, University of London, 1989), 1:xxiv–liii. "Alhacen" comes from his cognomen al-Hasan, although there is some possibility that Ibn al-Haytham was not one person, but two. See A. I. Sabra, "One Ibn al-Haytham or Two? An Exercise in Reading the Bio-Bibliographic Sources," *Zeitschrift für Geschichte der Arabisch-Islamischen Wissenschaften* 12 (1998): 1–50; for the opposite view, see Roshdi Rashed, ed., *Les mathématiques infinitésimales du IXe au XIe siècle*, vol. 2 (Al-Furquan, 1993), 1–19.

30. A. Mark Smith, *From Sight to Light: The Passage from Ancient to Modern Optics* (University of Chicago Press, 2015), 181–83.

31. Abdelhamid I. Sabra, "Ibn al-Haytham," in *Dictionary of Scientific Biography*, ed. Charles Coulston Gillispie (Charles Scribner's Sons, 1972), 189–210, 190.

32. "Omnia autem ea que diximus sic possunt experiri experiento cum quo veniet certificatio." Alhacen, *De aspectibus*, 2.25, in *Alhacen's Theory of Visual Perception: A Critical Edition, with English Translation and Commentary, of the First Three Books of Alhacen's 'De aspectibus,' the Medieval Latin Version of Ibn al-Haytham's Kitāb al-Manāẓir*, ed. and trans. A. Mark Smith, 2 vols. (American Philosophical Society, 2001), 1:263, English translation 2:573.

33. Sabra, *Kitāb al-manāẓir*, 3–4; Smith, ed., *Alhacen's Theory of Visual Perception*, xv–xxii, xxviii–xxxii; Sabra, "Ibn al-Haytham's Revolutionary Project in Optics: The Achievement and the Obstacle," in *The Enterprise of Science in Islam: New Perspectives*, ed. Jan. P. Hogendijk and Sabra (MIT Press, 2003), 85–118; Kheirandish, "The Mixed Mathematical Sciences," 98–99.

34. Smith, *From Sight to Light*, 225.

35. A. I. Sabra, "The 'Commentary' That Saved the Text: The Hazardous Journey of Ibn al-Haytham's Arabic *Optics*," *Early Science and Medicine* 12, no. 2 (2007): 117–33.

36. Smith, *From Sight to Light*, 14–16.

37. "Haec autem scientia non est adhuc lecta Parisius, nec apud Latinos, nisi bis Oxoniae in Anglia . . ." Bacon, *Opus tertium*, 1.11, ed. Nikolaus Egel (Felix Meiner, 2020), 76. On the dating of *De aspectibus* to the 1240s, see Hackett, "From *Sapientes antiqui*," 125.

38. Lindberg, "Roger Bacon on Light, Vision and the Universal Emanation of Force," in *Roger Bacon and the Sciences*, ed. Hackett, 243–75, esp. 245–50; David C. Lindberg, *Roger Bacon and the Origins of Perspectiva in the Middle Ages: A Critical Edition and English Translation of Bacon's Perspectiva with Introduction and Notes* (Clarendon Press, 1996), lxxxiii–lxxxvi.

39. Nader el-Bizri, "On Vision: Roger Bacon's *Opus maius* (V. 1–2) from the Perspective of Alhazen's *Optics* (1–2)," in *The Sciences of Roger Bacon's* Opus maius, ed. Meagan S. Allen (Routledge, 2025), 94–108; Katherine H. Tachau, *Vision and Certitude in the Age of Ockham: Optics, Epistemology, and the Foundation of Semantics, 1250–1345* (Brill, 1988), 3–19.

40. Mattia Mantovani, "The First of Natural Sciences: Roger Bacon on *Perspectiva* and Human Knowledge," *Vivarium* 59 (2021): 186–214.

41. Smith, *From Sight to Light*, 261–62.

42. On Bacon's repeated citations of Alhacen in *Opus maius*, see, for example, 5.3, in Roger Bacon, *Opus majus*, ed. John Henry Bridges, 3 vols. (Clarendon Press, 1897–1900), 2.137–41.

43. See under the *jidthr* ʿ.b.r in Hans Wehr, *Dictionary of Modern Written Arabic*, ed. J. Milton Cowan (Spoken Language Services, 1976); J. G. Hava, ed., *Al-Faraid Arabic-English Dictionary* (Catholic Press, 1964). The term appears often in religious texts, especially in the title, where it has a more layered meaning than simply "contemplation," (the closest English term) to mean active consideration and internal reflection leading to change. See, for example, Usāma ibn Munqidh, *Kitāb al- Iʿtibār* (*The Book of Contemplation*), ed. and trans. by Paul M. Cobb (Smithsonian Libraries, 2008).

44. Sabra, *The Optics of Ibn al-Haytham*, 2:18–19; Alhacen, *De aspectibus*, 2.26, ed. and trans. Smith, 1:263, English translation 2:573. See also Hackett, "Roger Bacon on *Scientia Experimentalis*," 289–90; Graziella Federici Vescovini, "La Fortune de l'optique d'Ibn Al-Haitham: Le livre *De aspectibus* (*Kitāb al-manāẓzir*) dans le moyen âge Latin," *Archives internationales d'histoire des sciences* 40 (1990): 220–38; Saleh Beshara Omar, *Ibn al-Haytham's*

Optics: A Study of the Origins of Experimental Science (Bibliotheca Islamica, 1977). On the earlier history of "experiment" in Arabic optics, see Elaheh Kheirandish, "Footprints of 'Experiment' in Early Arabic Optics," *Early Science and Medicine* 14 (2009): 79–104.

45. Abū Maʿšar, *Abū Maʿšar on Historical Astrology: The Book of Religions and Dynasties (On the Great Conjunctions)*, ed. Keiji Yamamoto and Charles Burnett, 2 vols. (Brill, 2000). See also Jole Agrimi and Chiara Crisciani, "Albumazar nell'astrologia di Ruggero Bacone," *Acme* 25, no. 3 (1972): 315–38; H. Darrel Rutkin, "The 'Unnamed Master' Revisited," in *The Sciences of Roger Bacon's* Opus maius, ed. Meagan S. Allen (Routledge, 2025), 109–29.

46. Paolo Zambelli, *The Speculum Astronomiae and Its Enigma: Astrology, Theology, and Science in Albertus Magnus and His Contemporaries* (Springer Dordecht, 1992); Laura Ackerman Smoller, *History, Prophecy, and the Stars: The Christian Astrology of Pierre d'Ailly, 1350–1420* (Princeton University Press, 1994), 30–31; Hilary M. Carey, "Astrology and Antichrist in the Later Middle Ages," in *Time and Eternity: The Medieval Discourse*, ed. Gerhard Jaritz and Gerson Moreno-Riaño (Brepols, 2003), 515–35, 519–20.

47. Lemay, "Roger Bacon's Attitude Toward the Latin Translations," 47; Nikolaus Egel, "Roger Bacon's New Founding of the Sciences in the Thirteenth Century," in *The Philosophy and Science of Roger Bacon: Studies in Honour of Jeremiah Hackett*, ed. Nicola Polloni and Yael Kedar (Routledge, 2021), 125–40, 132; Jeremiah Hackett, "Roger Bacon on the Classification of the Sciences," in *Roger Bacon and the Sciences*, ed. Hackett, 49–66.

48. "Nam per viginti annos quibus specialiter laboravi in studio sapientiae, neglecto sensu vulgi, plus quam duo millia librarum ego posui in his, propter libros secretos, et experientias varias, et linguas, et instrumenta, et tabulas, et alia; tum ad quaerendum amicitias sapientum, tum propter instruendos adjutores in linguis, in figuris, in numeris, in tabulis, et instrumentis, et multis aliis." Bacon, *Opus tertium*, 1.17, ed Egel, 120.

49. *De multiplicatione specierum* and *De speculis comburentibus*, in David C. Lindberg, ed., *Roger Bacon's Philosophy of Nature: A Critical Edition, with English Translation, Introduction and Notes, of De Multiplicatione Specierum and De Speculis Comburentibus* (Clarendon Press, 1983).

50. Peter Perigrinus of Maricourt, "Epistola de magnete," in *Petrus Peregrinus Opera*, ed. Loris Sturlese and Ron B. Thomson (Scuola Normale Superiore, 1995); in English, Peter Peregrinus of Maricourt, *The Letter of Petrus Peregrinus on the Magnet, A.D. 1269*, trans. Fr. Arnold (McGraw, 1904). On magnetism in medieval Latin culture before Peter, see Julian A. Smith, "Precursors to Peregrinus: The Early History of Magnetism and the Mariner's Compass in Europe," *Journal of Medieval History* 18, no. 1 (1992): 21–74. On Peter's work for the Angevin crown, see Pietro Egidi, ed., *Codice diplomatico dei Saraceni di Lucera* (L. Pierro et Figlio, 1917).

51. José Luis Rivera, "Pierre de Maricourt," in *A Companion to Philosophy in the Middle Ages*, ed. Jorge J. E. Garcia and Timothy B. Noone (Blackwell, 2003), 538–39; Andreas Kleinert, "Le moteur magnéto-mécanique de Pierre de Maricourt: Comptes-rendus du séminaire; Origine des idées scientifiques, ruptures et continuités," *Centre Commun d'Histoire des Sciences et d'Epistémologie de Lille* 1 (2004–5): 22–34; P. Erhard Schlund, "Petrus Peregrinus von Maricourt: Sein Leben und seine Schriften; Ein Beitrag zur Roger Bacon-forschung," *Archivum Fransciscanum Historicum* 4 (1911): 436–55; Robert Halleux, "Entre philosophie naturelle et savoir d'ingénieur: *L'Epistola de magnete* de Pierre de Maricourt," *Archives internationals d'Histoire des sciences* 56, no. 156–57 (2006): 3–17.

52. Crombie suggests that the relationship between Bacon and Peter was that of student to master, although there is no direct evidence for this. Crombie, *Robert Grosseteste and the*

Origins of Experimental Science, 204–10. See also Molland, "Roger Bacon's Knowledge of Mathematics," in *Roger Bacon and the Sciences*, ed. Hackett, 152–53; Williams, "Roger Bacon in Context," 129–32.

53. "Et ideo quod alii caecutientes nituntur videre, ut vespertilio lucem solis in crepusculo, ipse in pleno fulgore contemplator, propter hoc quod est dominus experimentorum; et ideo scit naturalia per experientiam, et medicinalia, et alkimistica, et omnia tam coelestia quam inferiora; immo verecundatur si aliquis laicus, vel vetula, vel miles, vel rusticus de rure sciat quae ipse ignorant. Unde omnia opera fundentium metalla, et quae operantur auro, et argento, et ceteris metallis, et omnibus mineralibus, ipse rimatus est; et omnia quae ad militiam, et ad arma, et ad venationes ipse novit; omnia quae ad agriculturam, et ad mensuras terrarium et opera rusticorum, examinavit; etiam experimenta vetularum et sortilegia, et carmina earum et omnium magicorum consideravit; et similiter omnium joculatorum illusions et ingenia; ut nihil quod sciri debeat lateat ipsum, et quatenus omnia falsa et magica sciat reprobare." Bacon, *Opus tertium*, 1.13, ed. Egel, 94–96.

54. "Sed nondum certificatum est de fractione per hujusmodi sanguinem, quanquam elaboratum est ad hoc; est sine illo sanguine potest frangi de facili. Hoc enim vidi oculis meis, et necesse est hoc, quia gemmae non possunt sculpi nisi per fragmenta hujus lapidis." Bacon, *Opus majus*, 6.1, ed. Bridges, 2:168.

55. "Sed non est ita, quia castor habet ea sub pectore, et tam mas quam femina hujusmodi testes producit." Bacon, *Opus majus*, 6.1, ed. Bridges, 2:168.

56. "Deinde vulgatum est, quod aqua calida citius congelatur qam frigida in vasis, et arguitur ad hoc quod contrarium excitatur per contrarium, sicut inimici sibi obviantes. Sed certum est quod aquia frigida citius congelatur experienti." Bacon, *Opus majus*, 6.1, ed. Bridges, 2:169.

57. For example, *Opus majus*, 1.4, ed. Bridges, 1:9–10.

58. "Aristoteles auctoritate Alexandri misit duo millia hominum per diversa loca mundi ut experientur omnia quae sunt in superficie terrae, sicut Plinius testator in Naturalibus." Bacon, *Opus majus*, 6.1, ed. Bridges, 2:169.

59. Truitt, "Knowledge and Power."

60. A late antique text that purported to be a copy of a letter that Alexander had sent to Aristotle, describing the marvels he had seen and the novel experiences he had had, circulated widely from the fourth century onward. See W. Walther Boer, ed., *Epistola Alexandri Macedonis ad Aristotelem magistrum suum de itinere suo et de situ Indiae*, Beiträge zur klassischen Philologie (Hain, 1980); this work has been edited and translated into English by Lloyd L. Gunderson as *Alexander's Letter to Aristotle About India* (Hain, 1980); Steven J. Williams, "Two Independent Textual Traditions? The Pseudo-Aristotelian *Secret of Secrets* and the Alexander Legend," in *Trajectoires européenes du Secretum secretorum du Pseudo-Aristote (XIIIe-XVIe siècle)*, ed. Catherine Gaullier-Bougassas, Margaret Bridges, and Jean-Yves Tilliette (Brepols, 2015), 29, 33–49.

61. Williams, "Two Independent Textual Traditions?," 27–54. On the mirror-for-princes genre in the Latin Middle Ages, see Derek Pearsall, "Hoccleve's *Regiment of Princes*: The Poetics of Royal Self-Representation," *Speculum* 69, no. 2 (1994): 386–410.

62. On the emergence of the text through accretion, see Mahmoud Manzalaoui, "The Pseudo-Aristotelian *Kitāb Sirr al-asrār*: Facts and Problems," *Oriens* 23–24 (1974): 147–257; Mario Grignaschi, "L'origine et les métamorphoses du *Sirr al-asrār*," *Archives d'histoire doctrinale et littéraire du moyen âge* 43 (1976): 7–112; Grignaschi, "Remarques sur la formation et l'interprétation du *Sirr al-asrār*," in *Pseudo-Aristotle, The Secret of Secrets: Sources and Influences*,

ed. W. F. Ryan and Charles B. Schmitt (The Warburg Institute, University of London, 1982), 3–33. On the possibility of an Indian origin for the text, see Stig Wikander, "De l'Inde à l'Espagne: L'origine de la *Poridat de las poridades*," *Actas IV Congresso de estudos árabes e islâmicos* (Brill, 1971), 267–69.

63. Mario Grignaschi, "Les *Rasā'il 'Aristātālīsa 'ilā-l-Iskandar* de Sālim Abū-l-'Alā et l'activité Culturelle a l'époque omayyade," *Bulletin d'études orientales de Damas* 19 (1965): 7–83; Mario Grignaschi, "La figure d'Alexandre chez les Arabes et sa genèse," *Arabic Sciences and Philosophy* 3, no. 2 (1993): 205–34.

64. Yahyā ibn al-Bitrīq (fl. 815) was a well-known translator of Aristotle's works in the early 'Abbasid period, but there is reason to doubt that he translated this text. See D. M. Dunlop, "The Translations of al-Bitrīq and Yahyā (Yuhannā) b. al-Batrīq," *Journal of the Royal Asiatic Society of Great Britain and Ireland* 91, no. 3–4 (1959): 140–50; Casimir Petraitis, *The Arabic Version of Aristotle's "Meteorology"* (Dar el-Machreq / Librarie orientale, 1967); Gerhard Endress, "The Circle of al-Kindī: Early Arabic Translations from the Greek and the Rise of Islamic Philosophy," in *The Ancient Tradition in Christian and Islamic Hellenism: Studies on the Transmission of Greek Philosophy and Sciences Dedicated to H. J. Drossaart Lulofs on His Ninetieth Birthday*, ed. Gerhard Endress and Remke Kruk (Brill, 1997), 43–76.

65. Dimitri Gutas, "Ethische Schriften im Islam," in *Orientalisches Mittelalter*, ed. Wolfhart Heinrichs (AULA Verlag, 1990), 346–65; Philip F. Kennedy, ed., *On Fiction and Adab in Medieval Arabic Literature* (Harrassowitz Verlag, 2005).

66. Steven J. Williams, *The Secret of Secrets: The Scholarly Career of a Pseudo-Aristotelian Text in the Latin Middle Ages* (University of Michigan Press, 2003), 60–94.

67. Williams, "The Early Circulation of the Pseudo-Aristotelian *Secret of Secrets* in the West: The Papal and Imperial Courts," in *Le scienze alla corte di Federico II*, Micrologus 2 (SISMEL Edizioni del Galluzzo, 1994): 127–44, esp. 132–39. On natural knowledge and philosophical inquiry at the imperial court, see Antonio Thiery, "Federico II e la conoscenza scientifica," in *Intellectual Life at the Court of Frederick II Hohenstaufen*, ed. William Tronzo (National Gallery of Art, 1994), 273–92.

68. Williams, *The Secret of Secrets*, 7–30.

69. "Sicut exemplaria quatuor que nunc inveni Oxonie non habuerunt illa, nec similter multa alia, set Parisus habui exemplaria perfecta," Bacon, *Opera hactenus inedita*, fasc. 5, *Secretum secretorum cum glossis et notulis*, ed. Steele, 39. Although Robert Steele, editor of Bacon's text, held that Bacon's edition of *Secretum secretorum* dates from the 1240s, more recent scholarship convincingly argues from internal evidence that, although Bacon first encountered *Secretum secretorum* in Paris during this period, he did not complete his edition of the text, along with his notes and introduction, until the 1270s. See Williams, "Roger Bacon and His Edition of the Pseudo-Aristotelian *Secretum secretorum*"; Williams, "The Early Circulation of the Pseudo-Aristotelian *Secret of Secrets*." On the debate over the authorship of *Secretum secretorum*, see Williams, "Defining the *Corpus Aristotelicum*: Scholastic Awareness of Aristotelian Spuria in the High Middle Ages," *Journal of the Warburg and Courtauld Institutes* 58 (1995): 29–51.

70. Jeremiah Hackett, "Roger Bacon, Aristotle, and the Parisian Condemnations of 1270, 1277," *Vivarium* 35, no. 2 (1997): 283–314.

71. "De istis scienciis naturalibus que vocari possunt inproprie geomancia, ydromancia, aerimancia, piromancia, que sunt vere partes philosophie, intendit Aristotiles in hoc libro

set translator non habuit in Latino nomina propria istis scienciis, ideo accepit nomina scienciarum magicarum que sunt similes aliquibus veris scienciis." Bacon, *Opera hactenus inedita*, fasc. 5, *Secretum secretorum cum glossis et notulis*, ed. Steele, 12.

72. See Bacon's commentary in *Opera hactenus inedita*, fasc. 5, *Secretum secretorum cum glossis et notulis*, ed. Steele, 38–39; Williams, *The Secret of Secrets*. See also Anthony Grafton, *Magus: The Art of Magic from Faustus to Agrippa* (Belknap Press of Harvard University Press 2023), 36–37.

73. See Bacon's commentary in *Opera hactenus inedita*, fasc. 5, *Secretum secretorum cum glossis et notulis*, ed. Steele, 8; Zachary A. Matus, *Franciscans and the Elixir of Life: Religion and Science in the Later Middle Ages* (University of Pennsylvania Press, 2017), 49.

74. There is no evidence that his edition reached the intended recipient. Williams, "Roger Bacon and His Edition of the Pseudo-Aristotelian *Secretum secretorum*," 65–67.

75. "Vestrae petitioni respondeo digilitenter." Bacon, *Epistola de secretis operibus artis et naturae*, 523. It is also possible that Bacon merely used this as a framing device for his work.

76. "Possunt enim sic figurari perspicua ut longissimi posita appareant proinquissima, et e converso; ita quod ex incredibili distantia legeremus minutissimas literas, et numeraremus res quascunque parvas." Bacon, *Epistola de secretis operibus artis et naturae*, 534.

77. Vincent Ilardi, *Renaissance Vision from Spectacles to Telescopes* (American Philosophical Society, 2007), 3–49; Smith, *From Sight to Light*, 270.

78. "Puer vero Johannes portavit crystallum sphaericum ad experiendum; et instruxi cum in demonstratione et figuratione huius rei occultae." Bacon, *Opus tertium*, 1.32, ed. Egel, 228.

79. Bacon, *Epistola de secretis operibus artis et naturae*, 536–38.

80. "Et primo per figuram et rationem solius artis. Nam instrumenta navigandi possunt fieri sine hominibus remigantibus, ut naves maxime, fluviales et marinae, ferantur unico homine regente, majori velocitate quam si plenae essent hominibus. Item currus possunt fieri ut sine animali moveantur cum impetus inaestimabili; ut aestimamus currus falcati fuisse, quibus antiquitus pugnabatur." Bacon, *Epistola de secretis operibus artis et naturae*, 533. On Bacon and the mechanical arts, see Elspeth Whitney, "The Artes Mechanicae, Craftsmanship and the Moral Value of Technology," in *Design and Production in Medieval and Early Modern Europe*, ed. Nancy van Deusen (Institute of Medieval Music, 1998), 75–87.

81. "Ut pontes ultra flumina sine columna, vel aliquo sustentaculo, et machinationes, et ingenia inaudita." Bacon, *Epistola de secretis operibus artis et naturae*, 533.

82. "Item instrumentum, parvum in quantitate ad elevandum et deprimendum pondera quasi infinta, quo nihil utilius est in casu. Nam per instrumentum altitudinis trium digitorum, et latitudinis eorundem, et minoris quantitates, posset homo seipsum et socios suos ab omni periculo carceris eripere, et elevare, et descendere." Bacon, *Epistola de secretis operibus artis et naturae*, 533.

83. "Posset etiam de facili fieri instrumentum quo unus homo traheret ad se mille homines per violentiam, mala eorum voluntate; et sic de rebus aliis attrahendis." Bacon, *Epistola de secretis operibus artis et naturae*, 533–35, quote on 533.

84. "Et occulta videantur manifesta." Bacon, *Epistola de secretis operibus artis et naturae*, 534–35.

85. For an overview of the scope of the Alexander tradition, see Richard Stoneman, *Alexander the Great: A Life in Legend* (Yale University Press, 2008). On Alexander's *pothos*, his longing for exploration and discovery, see Victor Ehrenberg, *Alexander and the Greeks*

(Blackwell, 1938), 52–61. On Alexander's inventiveness, see Jonathan Morton, "*Engin*: Creativity, Invention, and *Knowledge* in the Medieval Romance Tradition of Alexander the Great," *Romanic Review* 111, no. 2 (2020): 205–24.

86. Alain Rey, ed., *Dictionnaire historique de la langue française*, 2 vols. (Le Robert, 1992), s.v.v. "ingénieur," "engin"; Charles du Fresne Du Cange, ed., *Glossarium mediae et infimae latinitatis*, 10 vols. (Paris, 1937–38), s.v. "ingenium"; *Dictionnaire de l'ancienne langue française et tous ses dialects du IXe au XVe siècle*, ed. Frédéric Godefroy, 8 vols. (Slatkine Reprints, 1982), s.v.v. "engignart," "engigne," "engigneor." Cf. *OED*, s.v.v. "engine," "ingenious."

87. Thomas of Kent, "Hardiz estoit e conqueranz, sages e enginus," in *The Anglo-Norman Alexander (Le Roman de Toute Chevalerie)*, ed. Brian Foster with Ian Short, 2 vols. (Anglo-Norman Text Society, 1976), 1.7, l.30. This text is available with a modern French facing page translation: Alexandre de Paris, *Le roman d'Alexandre ou le roman de toute chevalerie*, trans. Catherine Gaullier-Bougassas and Laurence Harf-Lancner (Honoré Champion, 2003). Henry II was a student of Chartrian natural philosopher William of Conches in his youth; on the intellectual culture of the Plantagenet court, see Francine Mora-Lebrun, *"Metre en romanz": Les romans d'Antiquité du XIIème siècle et leur postérité (XIIIème-XIVème siècle)* (Honoré Champion, 2008), 53–86; Catherine Gaullier-Bougassas, ed., *La fascination pour Alexandre le Grand dans les littératures européenes (Xème-XVIème siècle)*, 4 vols. (Brepols, 2014), 2.794–98 and 3.1334–45.

88. Catherine Gaullier-Bougassas, "Savoir scientifique et 'roman historique': L'Alexandre anglo-normand de Thomas de Kent," in *Savoirs et fiction au Moyen Âge et à la Renaissance*, ed. Dominique Boutet and Joëlle Ducos (Presses de l'université de Paris-Sorbonne, 2015), 143–59.

89. Alexander of Paris, *The Medieval French Roman d'Alexandre*, vol. 2, *Version of Alexandre de Paris*, ed. Edward C. Armstrong, Douglas L. Buffum, Bateman Edwards, and L. F. H. Lowe (Princeton University Press, 1937), branche 1, lines 62–130, 2–3 and branche 3, lines 407–52, 152–55. Available with facing page translation into Modern French, although lacking part of the second *branche*, Alexandre de Paris, *Le roman d'Alexandre*, trans. Laurence Harf-Lancner (Livre de poche, 1994).

90. Alexandre de Paris, *Le roman d'Alexandre*, branche 2, lines 1904–2006, 115–18.

91. The alliterative verse Middle English "B" version of the Alexander romance (Bodl. MS Bodley 264) contains letters believed to be exchanged between Alexander and Dindimus, ruler of an ascetic population of philosophers in India. Omar Khalaf, ed., *Alexander and Dindimus* (Universitätsverlag Winter, 2017).

92. "Cogitavi cum amicis meis, ut instruerem tale ingeium, quatenus ascenderem caelum et viderem, si est hot caelum, quod videmus. Preparavi ingenium, ubi sederem, et apprehendi grifas et liquid eas cum catenas, et psui vectes ante eos et in summitate eorum cibaria illorum et ceperunt ascendere celum." Leo, "Die interpolierte Historia de preliis" 3.27, in *Der Alexanderroman des Archipresbyters Leo*, ed. Friedrich Pfister (C. Winter, 1913), 126.

93. "Et la voie du ciel refu par lui tentee / Quant la chaiere d'or en fu lassu portee / Par les quatre grifons, a qui fu acouplee; / Et fu d'astronomie sa pensee enluminee, / Que de toutes estoiles connut la compassee." Alexandre de Paris, *Le roman d'Alexandre*, branche 1, lines 71–77, 2.

94. Ingham, *The Medieval New*, 50–55; E. R. Truitt, "'Trei poëte, sages dotors, qui mout sorent di nigromance': Knowledge and Automata in Twelfth-Century French Literature," *Configurations* 12, no. 2 (2005): 167–93; Morton, "*Engin*"; Brian Stock, "The Self and

Literary Experience in Late Antiquity and the Middle Ages," *New Literary History* 25, no. 4 (1994): 839–52.

95. George Cary, *The Medieval Alexander*, ed. David J. A. Ross (Cambridge University Press, 1956), 46; Donald Maddox and Sara Sturm-Maddox, eds., *The Medieval French Alexander* (State University of New York Press, 2002); David J. A. Ross, *Studies in the Alexander Romance* (Pindar Press, 1985); Christiane Raynaud, "Aristote dans les enluminures du XIIIe siècle," in *L'unité de la culture européenne au Moyen Âge*, ed. Danielle Buschinger and Wolfgang Spiewok (Reineke Verlag, 1994).

96. "Item possunt fieri instrumenta volandi, ut homo sedeat in medio instruenti revolvens aliquod ingenium, per quod alae artificialiter compositae aerem verberent, ad modum avis volantis." Bacon, *Epistola de secretis operibus artis et naturae*, 533.

97. "Possunt etiam instrumenta fiere abmulandi in mari, vel fluminibus, usque ad fundum absque periculo corporali. Nam Alexander magnus his usus est, ut secreta maris videret, secundum quod Ethicus narrat astronomus." Bacon, *Epistola de secretis operibus artis et naturae*, 533.

98. "Sic enim aestimatur Julius Caesar super littus maris in Galliis, deprehendisse per ingentia specula dispositionem et situm castrorum et civitatum Britanniae majoris." Bacon, *Epistola de secretis operibus artis et naturae*, 534.

99. We see examples of this in both first-person accounts and in literary sources. In his mid-tenth-century account of his diplomatic embassy to Constantinople on behalf of Berengar II, King of Italy, in 949, Liudprand of Cremona described the levitating Throne of Solomon in terms of working like a wine press (although he had no explanation for the mechanical animals that surrounded the throne). And numerous romances describe metal people in familiar terms: a "maiden of copper," or "youths made of gold." See Liudprand of Cremona, *The Embassy to Constantinople and Other Writings*, ed. John Julius Norwich, trans. F. A. Wright (J. M. Dent, 1993), 153; E. R. Truitt, *Medieval Robots: Mechanism, Magic, Nature, and Art* (University of Pennsylvania Press, 2015), 22–24, 47–49.

100. Douglas Kelly, *The Art of Medieval French Romance* (University of Wisconsin Press, 1992), esp. 15–19.

101. Excellent on this point, though focused on other authorities that Bacon invoked, is David C. Lindberg, "Science as Handmaiden: Roger Bacon and the Patristic Tradition," *Isis* 78, no. 4 (1987): 518–36.

102. I rely here on the interpretation given by Patricia Ingham, in *The Medieval New*, 58–60.

103. "Et infinita quasi talia fieri possunt . . . et machinations, et ingenia inaudita." Bacon, *Epistola de secretis operibus artis et naturae*, 533.

104. "Et postquam huiusmodi perspexi, nihil mihi est difficile ad credendum, quando bene considero, nec in divinis, sicut nec in humanis." Bacon, *Epistola de secretis operibus artis et naturae*, 533.

CHAPTER 3

Note to epigraph. Frier Bacon, *His Discovery of the Miracles of Art, Nature, and Magick, Faithfully Translated out of Dr. Dee's Own Copy by T. M., and Never Before in English* (Printed for Simon Miller, 1659), A4v. "Discern" here has a wide range of meanings, including distinguishing between divine ecstasy and demonic influence. See *OED*, s.v. "discern"; Nancy Caciola, *Discerning Spirits: Divine and Demonic Possession in the Middle Ages* (Cornell University Press, 2003).

1."Et sicut in visione corporali, sic in spirituali oportet esse. Et hic potest fieri multiplex comparatio. Nam rectitude visionis Deo debetur; declination a rectitudine per fractionem, que debilior est, angelice nature convenit; reflexive visio, que est debilior, homini potest assignari." Bacon, *Perspectiva*, in *Roger Bacon and the Origins of Perspectiva in the Middle Ages: A Critical Edition and English Translation of Bacon's Perspectiva with Introduction and Notes*, ed. David C. Lindberg (Clarendon Press, 1996), 328–39. See also A. Mark Smith, *From Sight to Light: The Passage from Ancient to Modern Optics* (University of Chicago Press, 2015), 270–71.

2. Smith, *From Sight to Light*, 273–75; David C. Lindberg, "Lines of Influence in Thirteenth-Century Optics: Bacon, Witelo, and Pecham," *Speculum* 46, no. 1 (1971): 66–83, 73–75.

3. William of Moerbeke translated Hero of Alexandria's (d. 70 C.E.) work on mirrors, *Catoptrics*, into Latin for Witelo around 1269, and Witelo dedicated his work on optics, *Perspectiva*, to William. See Smith, *From Sight to Light*, 64n92; Agostino Paravicini Bagliani, *The Pope's Body*, trans. David S. Peterson (University of Chicago Press, 2000); 196–98; Agostino Paravicini Bagliani, ed., *Medicina e scienze della natura alla corte dei papi nel Duecento* (Centro Italiano di Studi sull'Alto Medioevo, 1991), 161–65.

4. On the similarities between *Opus maius* and Bonaventure's *Collationes*, see Jeremiah Hackett, "Aristotle, *Astrologia*, and Controversy at the University of Paris (1266–1274)," in *Learning Institutionalized: Teaching in the Medieval University*, ed. John van Engen (University of Notre Dame Press, 2000), 69–110, esp. 80–90. Amanda Power, *Roger Bacon and the Defence of Christendom* (Cambridge University Press, 2011), 72–73.

5. Smith, *From Sight to Light*, 272–75; David C. Lindberg, *Theories of Vision from Al-Kindi to Kepler* (University of Chicago Press, 1976), 117–20.

6. "Item frater Rogerius dictus Bachon Anglicus magis reali phylosophie studens quam scribende mirasa [*sic*] operates est in ea experimenta. Nam tante subtilitatis in naturali phylosophia extitit ut magis eius mirabilibus experimentis (quibus nulla verior scientia) quam scripture stilo aut doctrine verbo insistens, pontem ultra triginta miliaria longum naturaliter condempsans, super mare a terra firma in Anglie insulam per ipsum inde illuc cum tota sua comitiva transiens, aliquando fecit, ipsum post eius salubrem transitum similiter destruens, rarefaciendo naturaliter. Huiusmodi etiam cum semel in Anglia cuidam ioculatori, de ipso nugas facienti, cyrothecam quam secum portabat dedisset excuciendam ad quam dam, integramque in domo coram omnibus stabiliret [*sic*] columpnam, iseque eam excuciente primo percuteret ad eam eiusdem aeris simili condempsatione, sic fecit eius manum dextram qua sic excuciebat cum dicta cyrotheca absorber quod omnibus eidem viderentur immersum collumpne. De similibus et maioribus idem Rogerius mencionem facit in sua epistola ad papam Clementem quartum ad laudem Dei amen. Qualiter fratres minores fuerunt perspective capitulum quintum. . . . prephatus frater Rogerius Bachon, qui tante huiusmodi scientie plenitudine redundabat, ut delectatione experimentorum eius, obmissis doctrinis et scriptis componendis, aliquando in universitate Oxonie duo specula compsuit patientia, in quorum altero quilibet omni hora diei et noctis poterat accendere candela, in altero vero videre quid agebant homines in quantumcumque remotis constituti partibus. Et quia ad experimentum primi studentes plus stabant candelas accendendo quam in libris studendis, et in secundo multi, visis suis consanguineis et amicis mori, infirmari, aut aliter impediri, de universitate

recedentes stadium destruebant, eiusdem universitatis communi consilio utrumque est fractum." Peter of Trau, Franciscan, chronicle of deeds of Franciscans, Dubrovnik (1384–85), which appears in Oxford, Bodleian MS Canon, Misc. 525, fol. 202v–203v. Cited also in A. George Molland, "Roger Bacon as Magician," *Traditio* 30 (1974): 445–60, transcription at 446.

7. "Hic Generalis frater Hieronymus de multorum fratrum consilio condemnavit et reprobavit doctrinam fratris Rogerii Bachonis Anglici, sacrae theolgia magistri, continentem aliquas novitates suspectas, propter quas fuit idem Rogerius carceri condemnatus, praecipiendo omnibus fratribus, ut nullus illam teneret, ed ipsam vitaret ut per Ordinem reprobatam." *Analecta Franciscana*, vol. 3, *Chronica XXIV generalium Ordinis minorum* (Quaracchi, 1897), 360.

8. See Power, *Roger Bacon and the Defense of Christendom*, 15–26, for a thorough discussion of the charges leveled at Bacon after his death.

9. Amanda Power, "Seeking Remedies for Great Danger: Contemporary Appraisals of Roger Bacon's Expertise," in *Knowledge, Discipline, and Power in the Middle Ages: Essays in Honour of David Luscombe*, ed. Joseph Canning, Edmund J. King, and Martial Staub (Brill, 2011), 63–78, esp. 65–70.

10. Bert Roest, *Reading the Book of History: Intellectual Contexts and Educational Functions of Franciscan Historiography 1226–ca.1350* (Rijksuniversiteit, 1996).

11. David J. Collins, "Albertus, 'Magnus' or Magus? Magic, Natural Philosophy, and Religious Reform in the Late Middle Ages," *Renaissance Quarterly* 63, no. 1 (2010): 1–44; David J. Collins, *Disenchanting Albert the Great: The Life and Afterlife of a Medieval Magician* (Penn State University Press, 2024); E. R. Truitt, *Medieval Robots: Mechanism, Magic, Nature, and Art* (University of Pennsylvania Press, 2015), 91–94.

12. See also Patricia Clare Ingham, *The Medieval New: Ambivalence in an Age of Innovation* (University of Pennsylvania Press, 2015), 63–66.

13. James Hankins, *Virtue Politics: Soulcraft and Statecraft in Renaissance Italy* (Belknap Press of Harvard University Press, 2019), 153–73; Theodore E. Mommsen, "Petrarch's Conception of the 'Dark Ages,'" *Speculum* 17, no. 2 (1942): 226–42.

14. Ernst Robert Curtius, *European Literature and the Latin Middle Ages*, trans. Willard R. Trask (Pantheon Books, 1953), 251.

15. Jonathan Conant, *Staying Roman: Conquest and Identity in Africa and the Mediterranean, 439–700* (Cambridge University Press, 2012).

16. Curtius, *European Literature and the Latin Middle Ages*, 253–55. See also Rosamond McKitterick, *History and Memory in the Carolingian World* (Cambridge University Press, 2004).

17. Michael T. Clanchy, "*Moderni* in Education and Government in England," *Speculum* 50, no. 4 (1975): 671–88; Elisabeth Gössmann, *Antiqui und Moderni im Mittelalter: Eine geschichtliche Standortbestimmung* (Ferdinand Schöningh, 1974); Carol Symes, "When We Talk About Modernity," *American Historical Review* 116, no. 3 (June 2011): 715–26.

18. Mommsen, "Petrarch's Conception of the 'Dark Ages,'" 227.

19. Anthony Grafton and Megan Williams, *Christianity and the Transformation of the Book: Origen, Eusebius, and the Library of Caesarea* (Belknap Press of Harvard University Press, 2006), 133–77; Hervé Inglebert, "The Universal Chronicle in Antiquity and in the Middle Ages," in *Vehicles of Transmission, Translation, and Transformation in Medieval Textual Culture*, ed. Robert Wisnovsky, Faith Wallis, Jamie C. Fumo, and Carlos Fraenkel (Brepols, 2011), 75–101, 84–88.

20. François Hartog, *Regimes of Historicity: Presentism and Experiences of Time*, trans. Saskia Brown (Columbia University Press, 2015), 61–62.

21. Michael I. Allen, "Universal History 300–1000: Origins and Western Developments," in *Historiography in the Middle Ages*, ed. Deborah Mauskopf Deliyannis (Brill, 2006), 17–42; and in the same volume, Rolf Sprandel, "World Historiography in the Late Middle Ages," trans. Kristin E. Thomas, 157–80; and in the same volume, Deliyannis's "Introduction," 1–16; Gabrielle M. Spiegel, *Romancing the Past: The Rise of Vernacular Prose Historiography in Thirteenth-Century France* (University of California Press, 1993); Antonia Gransden, *Historical Writing in England, c. 550-c. 1307*, 2 vols. (Routledge, 1996 [Cornell University Press, 1974]). Other medieval thinkers also adopted Christian frameworks in different ways to understand human history. In the late twelfth century, the theologian Joachim of Fiore (1135–1202) divided time into the three ages of Father, Son, and Holy Spirit, the last a utopian age of divine communion. Although the movement of his followers that emerged was later condemned as heretical, his writings were influential within the Franciscan order in the first half of the thirteenth century. See Zachary A. Matus, *Franciscans and the Elixir of Life: Religion and Science in the Later Middle Ages* (University of Pennsylvania Press, 2017), 71–73.

22. Charles Homer Haskins, *The Renaissance of the Twelfth Century* (Harvard University Press, 1927); *Renaissance and Renewal in the Twelfth Century*, ed. Robert L. Benson, Giles Constable, and Carol D. Lanham (Harvard University Press, 1982).

23. Curtius, *European Literature and the Latin Middle Ages*, 29; J. G. A. Pocock, *Barbarism and Religion*, vol. 3, *The First Decline and Fall* (Cambridge University Press, 2003), 127–45.

24. Mommsen, "Petrarch's Conception of the 'Dark Ages,'" 230–36.

25. Mommsen, "Petrarch's Conception of the 'Dark Ages,'" 228. Cennino Cennini (ca. 1370–1440) a Florentine painter, wrote a practical handbook on painting, *Il Libro dell'Arte* (ca. 1390) in which he credited the earlier Florentine painter Giotto (ca. 1267–1337)—not coincidentally his teacher's teacher—with bringing painting up to date by returning to the principles of the ancients. See Christopher S. Wood, *A History of Art History* (Princeton University Press, 2019), 59–60.

26. Mommsen, "Petrarch's Conception of the 'Dark Ages,'" 240–41.

27. Giorgio Vasari, *The Lives of the Artists*, trans. Julia Conaway Bondanella and Peter Bondanella (Oxford University Press, 1998), 15. See also Philip Sohm, "Ordering History with Style: Giorgio Vasari on the Art of History," in *Antiquity and Its Interpreters*, ed. Alina Payne, Ann Kuttner, and Rebekah Smick (Cambridge University Press, 2000), 40–54.

28. Giorgio Vasari, *On Technique: Being the Introduction to the Three Arts of Design, Architecture, Sculpture and Painting*, trans. Louisa S. Maclehose (Dover, 1960 [J. M. Dent, 1907]), 83–84. See also Anne-Marie Sankovitch, "The Myth of the 'Myth of the Medieval': Gothic Architecture in Vasari's Rinascita and Panofsky's Renaissance," *RES: Anthropology and Aesthetics* 40 (2001): 29–50; Alina Payne, "Vasari, Architecture, and the Origins of Historicizing Art," *RES: Anthropology and Aesthetics* 40 (2001): 51–76.

29. Alexander Nagel and Christopher S. Wood, *Anachronic Renaissance* (Zone Books, 2010), 135–38.

30. Suzanne Conklin Akbari, *Seeing Through the Veil* (University of Toronto Press, 2004), 91–96. On the vernacularization—direct and indirect—of Ibn al-Haytham's work, see Smith, *From Sight to Light*, 278–98; Graziella Federici Vescovini, "Alhazen vulgarisé: Le De li

aspecti d'un manuscrit du Vatican (moitié du XIVe siècle) et le troisième Commentaire sur l'optique de Lorenzo Ghiberti," *Arabic Sciences and Philosophy* 8, no. 1 (1998): 67–96.

31. Akbari, *Seeing Through the Veil*, 138–46; Martin Kemp, *Visions of Heaven: Dante and the Art of Divine Light* (Lund Humphries, 2021).

32. Katherine H. Tachau, *Vision and Certitude in the Age of Ockham: Optics, Epistemology, and the Foundation of Semantics, 1250–1345* (Brill, 1988); Smith, *From Sight to Light*, 287–89.

33. Sarah M. Dillon, *Seeing Renaissance Glass: Art, Optics, and Glass of Early Modern Italy, 1250–1425* (Peter Lang, 2018), 132–34; Klaus Bergdolt, "Bacon und Giotto: Zum Einfluß der franziskanischen Naturphilosophie auf die bildende Kunst am Ende des 13. Jahrhunderts," *Medizinhistorisches Journal* 24 (1989): 25–41.

34. Scholars disagree on how much medieval treatises on optics influenced the depiction of linear perspective in Italian painting in the fifteenth century. See Smith, *From Sight to Light*, 299–303; Stuart Clark, *Vanities of the Eye: Vision in Early Modern European Culture* (Oxford University Press, 2007); Samuel Y. Edgerton Jr., *The Heritage of Giotto's Geometry: Art and Science on the Eve of the Scientific Revolution* (Cornell University Press, 1991); Samuel Y. Edgerton Jr., *The Mirror, the Window, and the Telescope: How Renaissance Linear Perspective Changed Our Vision of the Universe* (Cornell University Press, 2009); Hans Belting, *Baghdad and Florence: Renaissance Art and Arab Science*, trans. Deborah Lucas Schneider (Harvard University Press, 2011); Martin Kemp, *The Science of Art: Optical Themes in Western Art from Brunelleschi to Seurat* (Yale University Press, 1990); and Vescovini, "Alhazen vulgarisé." I say "(re)development" of linear perspective because artists in the Hellenic and Hellenistic periods also used linear perspective; see David Summers, *Vision, Reflection, and Desire in Western Painting* (University of North Carolina Press, 2007), 16–42.

35. Laura Ackerman Smoller, *History, Prophecy, and the Stars: The Christian Astrology of Pierre d'Ailly, 1350–1420* (Princeton University Press, 1994).

36. Anthony Grafton, *Magus: The Art of Magic from Faustus to Agrippa* (Belknap Press of Harvard University Press, 2023), 16. On natural magic and Italian humanism, see Frances A. Yates, *Giordano Bruno and the Hermetic Tradition* (University of Chicago Press, 1964).

37. Grafton, *Magus*, 66.

38. *OED*, s.v. "middle age" (*n.* & *adj.*).

39. Gregory B. Lyon, "Baudouin, Flacius, and the Plan for the Magdeburg Centuries," *Journal of the History of Ideas* 64, no. 2 (2003): 253–72.

40. Christoph Cellarius, *Historia universalis breviter ac perspicue exposita, in antiquam, et medii aevi ac novam divisa, cum notis perpetuis* (Jena, 1688).

41. Bacon, *His Discovery of the Miracles of Art, Nature, and Magick*, A4v.

42. Alexander Murray, "Should the Middle Ages Be Abolished?," *Essays in Medieval Studies* 21 (2004): 13–16.

43. Katharine Park, *Secrets of Women: Gender, Generation, and the Origins of Human Dissection* (Zone, 2006).

44. Meagan S. Allen, *Roger Bacon and the Incorruptible Human, 1220–1292: Alchemy, Pharmacology, and the Desire to Prolong Life*, Palgrave Studies in Medieval and Early Modern Medicine (Palgrave Macmillan, 2023).

45. Roger Bacon, *Opus majus*, ed. John Henry Bridges, 3 vols. (Oxford: Clarendon Press, 1897–1900), 6.2, 2.204–215; Roger Bacon, *Epistola de secretis operibus artis et naturae, et de nullitate magiae*, in *Fr. Rogeri Bacon opera quædam hactenus inedita*, ed. J. S. Brewer (Longman

and Green, 1859), 523–51, 537–42. William R. Newman, "An Overview of Roger Bacon's Alchemy," in *Roger Bacon and the Sciences*, ed. Jeremiah Hackett (Brill, 1997), 317–36; William R. Newman, "The Philosopher's Egg: Theory and Practice in the Alchemy of Roger Bacon," *Micrologus* 3 (1995): 75–101; Leah DeVun, *Prophecy, Alchemy, and the End of Time: John of Rupescissa in the Late Middle Ages* (Columbia University Press, 2009), 84–85.

46. Dorothea W. Singer, "Alchemical Writings Attributed to Roger Bacon," *Speculum* 7, no. 1 (1932): 80–86.

47. "Unde quicquid dicunt quod Salomon composuit hoc vel illud, aut alii sapientes, negandum est; quia non recipiuntur huusmodi libri auctoritate ecclesiae, nec a sapientibus, sed a seductoribus, qui mundum decipiunt. Nam et ipsi novos libros component, et novas adinventiones multiplicant, sicut scimus per experientiam; et tunc ut vehementius homines alliciant, praeponunt titulos famosos suis operibus, et eos magnis auctoribus ascribunt impudenter." Bacon, *Epistola de secretis operibus artis et naturae*, 526. See also Grafton, *Magus*, 26–27.

48. *De retardatione accidentum senectutis* is usually titled in English *The Cure of Old Age and the Preservation of Youth*, and was, for centuries, accepted as a Baconian text; however, more recent scholarship has proven that it is a spurious text. See Agostino Paravicini Bagliani, "Ruggero Bacone autore del *De retardatione accidenitum senectutis*?," *Studi medievali* 28, no. 2 (1987): 707–27; Paravicini Bagliani and Steven J. Williams, "Addendum," *Studi medievali* 28, no. 2 (1987): 727–28.

49. Roger Bacon, *Opus majus*, ed. Samuel Jebb (William Bowyer, 1733).

50. "Prestigiator ac magus necromanticus, non in virtute Dei, sed in operatione malorum spirituum Oxonii ad nasum eneum, scholasticorum domicilium, mirabilia magna fecisse traditur." John Bale, *Scriptorum illustrium maioris Britanniae . . . Catalogus* 1st ed. (John Overton, 1548), fol. 114v; "Acessit ei in Mathesi peritia incredibilis, sed absque Necromantia: quamuis ea a multis infametur" (2nd ed. [Basel, 1559], fol. 342). "Mathesis" refers to mathematical sciences, including astral divination. See also Wood's account of Bale's characterization of Bacon: Anthony Wood, *Survey of the Antiquities of the City of Oxford, Composed in 1661–6, by Anthony Wood*, ed. Andrew Clark, 3 vols. (Oxford Historical Society, 1889–99), 2:385.

51. Molland, "Roger Bacon as Magician," 447.

52. On eleventh and early twelfth-century suggestions that men learned in the *quadrivium* were practicing necromancers, see E. R. Truitt, "Celestial Divination and Arabic Science in Twelfth-Century England: The History of Gerbert of Aurillac's Talking Head," *Journal of the History of Ideas* 73 (2012): 201–22; on the same suggestion attached to philosophers of the thirteenth century, including Grosseteste and Albert the Great, see Truitt, *Medieval Robots*, 69–95.

53. Carol Z. Wiener, "The Beleaguered Isle: A Study of Elizabethan and Early Jacobean Anti-Catholicism," *Past and Present* 51 (1971), 27–62.

54. Gabriel Naudé, *Apologie pour tous les grandes personnages qui ont esté faussement soupçonnez de magie* (François Targa, 1625), 491–95; translated into English by John Davies and published as *The History of Magick by Way of Apology, for All the Wise Men Who Have Unjustly Been Reputed Magicians, from the Creation, to the Present Age* (London, 1657). See also Lauren Kassell, "'All Was This Land Full Fill'd of Faerie,' or Magic and the Past in Early Modern England," *Journal of the History of Ideas* 67, no. 1 (2006): 107–22.

55. Amanda Power, "A Mirror for Every Age: The Reputation of Roger Bacon," *English Historical Review* 121, no. 492 (2006): 657–92, 668.

56. Power, "A Mirror for Every Age," 663; Deanne Williams, "*Friar Bacon and Friar Bungay* and the Rhetoric of Temporality," in *Reading the Medieval in Early Modern England*, ed. Gordon A. McMullan and David Matthews (Cambridge University Press, 2007), 31–48.

57. Steven J. Williams, "Roger Bacon in Context: Empiricism in the High Middle Ages," in *Expertus sum: L'expérience par les sens dans la philosophie naturelle médiévale*, ed. Thomas Bénatouïl and Isabelle Draelants, Micrologus 40 (SISMEL Edizioni del Galluzzo, 2011), 123–44, 138–39.

58. Jennifer M. Rampling, *The Experimental Fire: Inventing English Alchemy, 1300–1700* (University of Chicago Press, 2020).

59. R. Ian McCallum, "Alchemical Scrolls Associated with George Ripley," in *Mystical Metal of Gold: Essays on Alchemy and Renaissance Culture*, ed. Stanton J. Linden (AMS Press, 2007), 161–88; Jonathan Hughes, "The Humanity of Thomas Charnock, an Elizabethan Alchemist," in *Mystical Metal of Gold: Essays on Alchemy and Renaissance Culture*, ed. Stanton J. Linden (AMS Press, 2007), 3–34, 5.

60. Thomas Digges, Leonard Digges, *An Arithmeticall Militare Treatise Named Stratioticos . . . Long Since Attempted by Leonard Digges, Gentleman: Augmented, Digested and Lately Finished by Thomas Digges, His Sonne* (Henrie Bynneman, 1579), 189–90.

61. See Robert Steele, "Roger Bacon and the State of Science in the Thirteenth Century," in *Studies in the History and Method of Science*, Charles Singer, vol. 2 (Clarendon Press, 1921), 121–50, 147; Molland, "Roger Bacon as Magician," 457–58.

62. Power, "A Mirror for Every Age: The Reputation of Roger Bacon," 671.

63. Rampling, *Experimental Fire*, 207.

64. Nicholas H. Clulee, "Astrology, Magic, and Optics: Facets of John Dee's Early Natural Philosophy," *Renaissance Quarterly* 30, no. 4 (1977): 632–80, 640n27 and 664n97.

65. Clulee, "Astrology, Magic, and Optics," 643–44.

66. Because the propagation of species could be described as occurring in straight lines, or rays, all optical phenomena could be represented as geometrical relationships conforming to geometrical rules. See David C. Lindberg, "Introduction," in *Roger Bacon and the Origins of Perspectiva in the Middle Ages*, esp. xxv–xciii; Clulee, "Astrology, Magic, and Optics," 663, 669; Nicholas H. Clulee, "John Dee's Mathematics and the Grading of Compound Qualities," *Ambix* 18 (1971): 178–211.

67. See John Dee, *Propaedeumata aphoristica* (Henricus Suttonus, 1558), sig. *iii^{r-v}, cited in Clulee, "Astrology, Magic, and Optics," 672.

68. Glyn Parry, *The Arch Conjuror of England: John Dee* (Yale University Press, 2011), 30–37.

69. Roger Bacon, *Epistolae fratris Rogerii Baconis, De secretis operibus artis et naturae, et de nullitate magiae / opera Iohannis Dee Londinensis e pluribus exemplaribus castigata olim, et ad sensum integrum restituta: Nunc vero a quodam veritatis amatore, in gratiam verae scientiae candidatorum foras emissa, cum notis quibusdam partim ipsius Johannis Dee, partim edentis* (Ex Bibliopolio Frobeniano, 1618).

70. "Siquidem utile genus eorum est, qui de theoriis non admodum solliciti, mechanica quadam subtilitate rerum inventarum extensiones prehendunt; qualis est Bacon." Francis Bacon, *Temporis partus masculus*, in *The Works of Francis Bacon*, vol. 3, *Philosophical Works*, ed. James Spedding, Robert Leslie Ellis, Douglas Denon Heath (Garret Press, 1968 [Longmans, 1857]), 3:534.

71. Jeremiah Hackett, "Roger Bacon on *Scientia Experimentalis*, in *Roger Bacon and the Sciences* ed. Hackett (Brill, 1997), 277–316, 277n1.

72. Roger Bacon, *The Cure of Old Age and the Preservation of Youth, by Roger Bacon, Translated out of Latin, with Annotations, and an Account of His Life and Writings*, ed. Richard Browne (Printed for Tho. Flesher . . . and Edward Evets . . . , 1683).

73. Thomas Birch, "March/April 1679," *The History of the Royal Society of London for Improving of Natural Knowledge, from Its First Rise in Which the Most Considerable of Those Papers Communicated to the Society, Which Have Hitherto Not Been Published, Are Inserted in Their Proper Order, as a Supplement to the Philosophical Transactions*, 4 vols. (A. Millar, 1757), 3:470–79. See also Stanton J. Linden, "Introduction," in *The Mirror of Alchimy: Composed by the Thrice-Famous and Learned Fryer, Roger Bachon*, ed. Stanton J. Linden (Garland Publishing, 1992), xliv; Power, "A Mirror for Every Age," 668–69.

74. Robert Hooke, "Lecture on the 'arithmatick of the Brachmans' and of a sort of 'Carolina cloath,'" 1689, CLP/20/79, London, The Royal Society.

75. Hooke, "Lecture on the 'arithmatick of the Brachmans.'"

76. Hooke, "Lecture on the 'arithmatick of the Brachmans.'"

77. My gratitude to Charles Burnett, Thony Christie, and Priya Nambrath for helping me contextualize this passage.

78. On Bacon as inventor of the telescope, see Thomas Birch, "July 1682," *The History of the Royal Society of London for Improving of Natural Knowledge, from Its First Rise in Which the Most Considerable of Those Papers Communicated to the Society, Which Have Hitherto Not Been Published, Are Inserted in Their Proper Order, as a Supplement to the Philosophical Transactions*, vol. 4 (A. Millar, 1757), 156–57; also see Robert Plot, *The Natural History of Oxford-Shire, Being an Essay Towards the Natural History of England* (Oxford, 1677), 214–21. On the status of innovation as a value in the sixteenth and seventeenth centuries, see Lorraine Daston and Katharine Park, "The Age of the New," in *Cambridge History of Science*, vol. 3, *Early Modern Science* (Cambridge University Press, 2006), 1–17.

79. John Aubrey, *Brief Lives with an Apparatus for the Lives of our English Mathematical Writers*, ed. Kate Bennett (Oxford University Press, 2015), 19.

80. Dag Nikolaus Hasse, *Success and Suppression: Arabic Sciences and Philosophy in the Renaissance* (Harvard University Press, 2016), 7–27.

81. Hasse, *Success and Suppression*, 69–133.

82. Hasse, *Success and Suppression*, 341–57; Craig Martin, "Rethinking Renaissance Averroism," *Intellectual History Review* 17, no. 1 (2007): 3–19; Craig Martin, *Subverting Aristotle: Religion, History, and Philosophy in Early Modern Science* (Johns Hopkins University Press, 2014).

83. Alexander Bevilacqua, *The Republic of Arabic Letters: Islam and the European Enlightenment* (Belknap Press of Harvard University Press, 2018), 44–135.

84. Bevilacqua, *The Republic of Arabic Letters*, 17–43.

85. Matthew Melvin-Koushki, "Taḥqīq vs. Taqlīd in the Renaissances of Western Early Modernity," *Philological Encounters* 3, no. 1–2 (2018): 193–249, 207–9.

86. Melvin-Koushki, "Taḥqīq vs. Taqlīd," 211.

87. Melvin-Koushki, "Taḥqīq vs. Taqlīd," 211.

88. Melvin-Koushki, "Taḥqīq vs. Taqlīd," 226. We also see evidence of the Orientalization of Arabic knowledge in the twelfth-century legends of Gerbert of Aurillac. See Truitt, "Celestial Divination and Arabic Science."

89. Hasse, *Success and Suppression*, 147–300.

90. Ingham, *The Medieval New*, 2.

CHAPTER 4

Notes to epigraphs. "Sunt enim non minus temporum quam regionum eremi et vastitates." Francis Bacon, *Novum organum*, 1.78, ed. Thomas Fowler (Clarendon Press, 1878), 272. William Whewell, *History of the Inductive Sciences, from the Earliest to the Present Time*, 3 vols., 3rd ed. (J. W. Parker, 1857 [1837]), 1:230–31.

1."Media mundi tempora, quod scientiarum segetem uberem aut laetam, infoelicia fuerunt. Neque enim causa est ut vel Arabum vel scholasticorum mentio fiat: qui per intermedia tempora scientias potius contriverunt numerosis tractatibus, quam pondus earum auxerunt." Bacon, *Novum organum*, 1.78, ed. Fowler, 272–73.

2. E. R. Truitt, "Making English Scientific: Chaucer, Translation and the Astrolabe," *Isis* 115, no. 4 (2024): 757–75, 760–62.

3. Lucien Fébvre, "How Jules Michelet Invented the Renaissance," in *A New Kind of History: from the writings of Febvre*, ed. Peter Burke, trans. K. Folca (Routledge and Kegan Paul, 1973), 258–67, 258.

4. Theodore E. Mommsen, "Petrarch's Conception of the 'Dark Ages,'" *Speculum* 17, no. 2 (1942): 226–42, 227–28; Catherine Brown, "In the Middle," *Journal of Medieval and Early Modern Studies* 30, no. 3 (2000): 547–74, 547–49; John H. Arnold, *What Is Medieval History?* (Polity, 2008), 5–9.

5. Carol Symes, "When We Talk About Modernity," *American Historical Review* 116, no. 3 (2011): 715–26, 721. For narratives of modernity that characterize the medieval period as necessarily un- or anti-modern, see Max Weber, "Science as Vocation," in *From Max Weber: Essays in Sociology*, eds. H. H. Gerth and C. Wright Mills (Oxford University Press, 1946); Stephen Greenblatt, *The Swerve: How the World Became Modern* (W. W. Norton, 2011); David Wooton, *The Invention of Science: A New History of the Scientific Revolution* (HarperCollins, 2015).

6. Symes, "When We Talk About Modernity," 716–17; Lee Patterson, *Literary Practice and Social Change in Britain, 1380–1530* (University of California Press, 1990), 1–14.

7. Judith M. Bennett, *History Matters: Patriarchy and the Challenge of Feminism* (University of Pennsylvania Press, 2006), 40–64; José Rabasa, "Decolonizing Medieval Mexico," in *Medievalisms the Postcolonial World: The Idea of "the Middle Ages" Outside Europe*, ed. Kathleen Davis and Nadia Altschul (Johns Hopkins University Press, 2009), 27–50.

8. Johannes Fabian, *Time and the Other: How Anthropology Makes Its Object* (Columbia University Press, 2014 [1983]). Other scholars have extended the "pre" to encompass the period before writing (prehistory) and the period before the emergence of *Homo sapiens*, arguing for ways of understanding the deep past that both are free of the vicelike grip of "modernity" yet also reliant on knowledge and knowledge-production emblematic of modernity. See Daniel Lord Smail and Andrew Shryock, "History and the 'Pre,'" *American Historical Review* 118, no. 3 (2013): 709–37; Daniel Lord Smail, *On Deep History and the Brain* (University of California Press, 2008). Elsewhere, Smail argues that all history is predicated on an idea of modernity—a present claim to newness and superiority, when compared to whatever came before. Daniel Lord Smail, "In the Grip of Sacred History," *American Historical Review* 110, no. 5 (2005): 1337–61.

9. Fabian, *Time and the Other*, 3; Brown, "In the Middle," 549–51.

10. John Dagenais and Margaret Rich Greer in "Decolonizing the Middle Ages: Introduction," in "Decolonizing the Middle Ages," special issue, *Journal of Medieval and*

Early Modern Studies 30, no. 3 (2000): 431–48, 431. See also Dipesh Chakrabarty, "Historicism and Its Supplements: A Note on a Predicament Shared by Medieval and Postcolonial Studies," in *Medievalisms in the Postcolonial World: The Idea of "the Middle Ages" Outside Europe*, ed. Kathleen Davis and Nadia Altschul (Johns Hopkins University Press, 2009), 109–19.

11. Michelle R. Warren, *Creole Medievalism: Colonial France and Joseph Bédier's Middle Ages* (University of Minnesota Press, 2011), 2–6.

12. Alexander Murray, "Should the Middle Ages Be Abolished?," *Essays in Medieval Studies* 21 (2004): 1–22, 2–4.

13. Dagenais and Greer "Decolonizing the Middle Ages," 431. See also Patrick J. Geary, *The Myth of Nations: The Medieval Origins of Europe* (Princeton University Press, 2002), 16–19.

14. Kathleen Davis, *Periodization and Sovereignty: How Ideas of Feudalism and Secularization Govern the Politics of Time* (University of Pennsylvania Press, 2008); John Ganim, *Medievalism and Orientalism* (Palgrave, 2005).

15. Edward W. Said, *Orientalism* (Vintage, 1979); Dipesh Chakrabarty, "Postcoloniality and the Artifice of History: Who Speaks for 'Indian' Pasts?," *Representations* 37 (1992): 1–26; Dipesh Chakrabarty, *Provincializing Europe: Postcolonial Thought and Historical Difference* (Princeton University Press, 2007 [2000]); Dipesh Chakrabarty "The Muddle of Modernity," *American Historical Review* 116, no. 3 (2011): 663–75, 663–64. See also Brown, "In the Middle," 549–50, although she argues also that medievalism and Orientalism are significantly different in their effects on the lives of people. The discourses of Orientalism and medievalism are context-dependent, as Karla Mallette has shown in *European Modernity and the Arab Mediterranean: Toward a New Philology and a Counter-Orientalism* (University of Pennsylvania Press, 2010). On the creation of the category of "early modernity" as a way of bridging the conceptual distance between "medieval" and "modern," see Randolph Starn, "The Early Modern Muddle," *Journal of Early Modern History* 6, no. 3 (2002): 296–307. Scholars of the medieval period have grappled with how and whether postcolonial theory can be applied to the period. See Gabrielle M. Spiegel, "Épater les médiévistes," *History and Theory* 39, no. 2 (2000): 243–50; Brown, "In the Middle"; Jeffrey Jerome Cohen, ed., *The Postcolonial Middle Ages* (Palgrave, 2000); Patricia Clare Ingham and Michelle R. Warren, eds., *Postcolonial Moves: Medieval Through Modern* (Palgrave Macmillan, 2003); Nadia Altschul, "Postcolonialism and the Study of the Middle Ages," *History Compass* 6, no. 2 (2008): 588–606.

16. See Matthew X. Vernon, *The Black Middle Ages: Race and the Construction of the Middle Ages* (Palgrave Macmillan, 2018).

17. Geary, *Myth of Nations*, 26–32.

18. Symes, "When We Talk About Modernity," 722.

19. Charles L. Eastlake, *A History of the Gothic Revival: An Attempt to Show How the Taste for Mediaeval Architecture, Which Lingered in England During the Two Last Centuries, Has Since Been Encouraged and Developed* (Longmans, Green, and Co., 1872), 112–31.

20. Scott's popularity in the United States, especially in the American South, was so extensive that Mark Twain could blame the medievalism of the Confederacy on interest in his work. See Mark Twain [Samuel Langhorne Clemens], *Life on the Mississippi* (James Osgood and Co., 1883), 416–19, 465–70.

21. William J. Ashworth, *The Trinity Circle: Anxiety, Intelligence, and Knowledge Creation in Nineteenth Century England* (University of Pittsburgh Press, 2021), 8.

22. Ashworth, *Trinity Circle*, 58.

23. Ashworth, *Trinity Circle*, 36.

24. Harvey W. Becher, "William Whewell's Odyssey: From Mathematics to Moral Philosophy," in *William Whewell: A Composite Portrait*, ed. Menachem Fisch and Simon Schaffer (Oxford University Press, 1991), 1–30, 4–6. William Whewell, *An Elementary Treatise on Mechanics* (J. Smith . . . for J. Deighton and Sons, 1819), 240–56; William Whewell, *The Mechanics of Engineering: Intended for Use in Universities and Colleges of Engineers* (John W. Parker, 1841), 84–111, all cited in Becher.

25. William Whewell, *Architectural Notes on German Churches, with Remarks on the Origin of Gothic Architecture*, 2nd ed. (J. and J. J. Deighton, 1835), 17.

26. Whewell, *Architectural Notes*, xxiv–xxv.

27. Whewell, *Architectural Notes*, 5.

28. Becher, "William Whewell's Odyssey," 6.

29. Edward Gillin, "Whewell and Architecture," in *William Whewell: Victorian Polymath*, ed. Lukas M. Verburgt (University of Pittsburgh Press, 2024), 61–80, 73–75.

30. Aleta Quinn, "William Whewell's Philosophy of Architecture and the Historicization of Biology," *Studies in History and Philosophy of Science Part C: Studies in History and Philosophy of Biological and Biomedical Sciences* 59 (2016): 11–19.

31. Ashworth, *Trinity Circle*, 70–71, 110; Richard Yeo, *Defining Science: William Whewell, Natural Knowledge, and Public Debate in Early Victorian Britain* (Cambridge University Press, 1993), 148–49; Geoffrey N. Cantor, "Between Rationalism and Romanticism: Whewell's Historiography of the Inductive Sciences," in *William Whewell: A Composite Portrait*, ed. Menachem Fisch and Simon Schaffer (Oxford University Press, 1991), 67–86, 75–77.

32. Philippa Levine, *The Amateur and the Professional: Antiquarians, Historians, and Archaeologists in Victorian England, 1838–1886* (Cambridge University Press, 1986).

33. Whewell, *History of the Inductive Sciences*, 1:254.

34. Cantor, "Between Rationalism and Romanticism," 69; Yeo, *Defining Science*, 156–60.

35. Laura J. Snyder, "Renovating the *Novum Organum*: Bacon, Whewell, and Induction," *Studies in History and Philosophy of Science* 30, no. 4 (1999): 531–57; Laura J. Snyder, "William Whewell," *The Stanford Encyclopedia of Philosophy*, ed. Edward N. Zalta, Spring 2019, https://plato.stanford.edu/archives/spr2019/entries/whewell/.

36. Laura J. Snyder, "Discoverers' Induction," *Philosophy of Science* 64, no. 4 (1997): 580–604.

37. See Ashworth, *Trinity Circle*; Laura J. Snyder, *Reforming Philosophy: A Victorian Debate on Science and Society* (University of Chicago Press, 2006).

38. Ashworth, *Trinity Circle*, 118–25.

39. On seventeenth-century Europe as the basis for modern science, and the Eurocentric ideology this historical narrative served, see Lorraine Daston, "The History of Science and the History of Knowledge," *KNOW: A Journal on the Formation of Knowledge* 1 (2017): 131–54, https://doi.org/10.1086/691678.

40. On Whewell and the Mathematics Tripos curriculum at Cambridge, see Ashworth, *Trinity Circle*, 106–7, 153.

41. Whewell, *History of the Inductive Sciences*, 1:180.

42. Whewell, *History of the Inductive Sciences*, 1:252–53.

43. Whewell, *History of the Inductive Sciences*, 1:254.

44. Whewell, *History of the Inductive Sciences*, 1:14.

45. Whewell, *History of the Inductive Sciences*, 1:vii; 1:181–82.

46. Whewell, *History of the Inductive Sciences*, 1:182.

47. Whewell, *History of the Inductive Sciences*, 1:182.

48. Whewell, *History of the Inductive Sciences*, 1:204.

49. Whewell, *History of the Inductive Sciences*, 1:245.

50. Whewell, *History of the Inductive Sciences*, 1:367.

51. Whewell, *History of the Inductive Sciences*, 1:367–70. (This is a reprint, included in the third edition of *History*, as part of the entry on Bacon in *Philosophy*.)

52. Whewell to Hugh James Rose, December 12, 1826, in Isaac Todhunter, ed., *William Whewell, D. D., Master of Trinity college, Cambridge: An Account of His Writings with Selections from His Literary and Scientific Correspondence*, 2 vols. (Macmillian, 1876), 2:77–79, and discussed in Ashworth, *Trinity Circle*, 68–69.

53. Cantor, "Between Rationalism and Romanticism," 70–71.

54. Whewell, *History of the Inductive Sciences*, 1:373.

55. Whewell, *History of the Inductive Sciences*, 1:211.

56. Whewell, *History of the Inductive Sciences*, 1:212.

57. Whewell, *History of the Inductive Sciences*, 1:230–31.

58. Whewell, *History of the Inductive Sciences*, 1:258–59.

59. Whewell, *History of the Inductive Sciences*, 1:212.

60. Whewell, *History of the Inductive Sciences*, 1:213. This characterization of Arabic authors in natural philosophy in the medieval period held sway from Whewell's time through most of the twentieth century. See Abdelhamid I. Sabra, "The Appropriation and Subsequent Naturalization of Greek Science in Medieval Islam: A Preliminary Statement," *History of Science* 25, no. 3 (1987): 223–43; Ahmed Ragab, "Translation and the Making of a Medical Archive: The Case of the Islamic Translation Movement," *Osiris* 37 (2022): 25–46.

61. Whewell, *History of the Inductive Sciences*, 1:259–60.

62. Matthew Melvin-Koushki, "Taḥqīq vs. Taqlīd in the Renaissances of Western Early Modernity," *Philological Encounters* 3, no. 1–2 (2018): 193–249.

63. Whewell conflated work written in Arabic with work written by Muslim scholars, although this conflation is at odds with the realities of textual production in the medieval period.

64. Whewell, *History of the Inductive Sciences*, 1:231.

65. William Whewell to Rev. Richard Jones, October 1837, in Todhunter, ed., *William Whewell*, 2:261. Emphasis in original.

66. William Whewell, "Letter to the Editor," *Edinburgh Review*, January 3, 1838.

67. William Whewell, research notes, Cambridge, Trinity College Library, R. 18.17 (13), fols. 61v–71r, quote at fol. 69v. These notes are also briefly discussed in Yeo, *Defining Science*, 153.

68. William Whewell, research notes, Cambridge, Trinity College Library, R. 18.17 (15), fols. 24r–27r, quote at fol. 25r.

69. William Whewell, research notes, Cambridge, Trinity College Library, R.18.10 (1), fol. 24. Every historical actor Whewell noted, with one exception, lived under Muslim political control and was considered Muslim and Arab by Whewell (regardless of the actual ethnic or religious background). The inclusion of Charles Martel, founder of the Carolingian dynasty and grandfather of Charlemagne, is notable for two reasons: (1) Charles Martel has no link to astral science (or the history of science more broadly) in any way; (2) he is

credited with the defeat of the Umayyads in 732 at the Battle of Tours and halting the Umayyad expansion of territory north of Iberia.

70. Augustus de Morgan to John Herschel, February 11, 1859, London, The Royal Society, HS/6/324.

71. Whewell, *History of the Inductive Sciences*, 1:231. The equation of the medieval past with the "uncivilized" parts of the globe (i.e., elsewhere than Europe and settler-occupied North America) remained widely applied for over a century after Whewell. See Chakrabarty, "Historicism and Its Supplements"; Norman Etherington, "Barbarians Ancient and Modern," *American Historical Review* 116 (2011): 31–57.

72. On Bacon's supposed confinement, see Whewell, *History of the Inductive Sciences*, 1:377.

73. Magda Teter, *Christian Supremacy: Reckoning with the Roots of Antisemitism and Racism* (Princeton University Press, 2023), 16–31; David Nirenberg, *Anti-Judaism: The Western Tradition* (Norton, 2013).

74. Anthony Grafton and Megan Williams, *Christianity and the Transformation of the Book: Origen, Eusebius, and the Library of Caesarea* (Belknap Press of Harvard University Press, 2006), 86–177. Christian sects continue to wrestle with how to relate to pre-Christian time in a way that is not necessarily supersessionary. See Kathleen Biddick, *The Typological Imaginary: Circumcision, Technology, History* (University of Pennsylvania Press, 2003), 7.

75. Beryl Smalley, *The Study of the Bible in the Middle Ages*, 3rd ed. (Blackwell, 1983 [1952]), esp. 214–52; Erich Auerbach, "Figura," first published in 1944 in *Neue Dantestudien*, and then translated and included in *Scenes from the Drama of European Literature* (University of Minnesota Press, 1994).

76. John Lowden, *The Making of the Bibles Moralisées*, 2 vols. (Penn State University Press, 2000); Sara Lipton, *Images of Intolerance: The Representation of Jews and Judaism in the Bible Moralisée* (University of California Press, 1999); Katherine H. Tachau, "God's Compass and *Vana Curiositas*: Scientific Study in the Old French *Bible Moralisée*," *Art Bulletin* 80, no. 1 (1998): 7–33.

77. On this point see Smail, "In the Grip of Sacred History"; Biddick, *The Typological Imaginary*.

78. This is the narrative presented in the most widely used account of medieval science, David C. Lindberg, *The Beginnings of Western Science: The European Scientific Tradition in Philosophical, Religious, and Institutional Context, Prehistory to A.D. 1450*, 2nd ed. (University of Chicago Press, 2007). For critiques of this view, see Sabra, "The Appropriation and Subsequent Naturalization of Greek Science in Medieval Islam"; Ahmed Ragab, "Translation and the Making of a Medical Archive."

79. Yeo, *Defining Science*, 168–69, 171.

80. Yeo, *Defining Science*, 5–6.

CHAPTER 5

Note to epigraph. Robert Adamson, *Roger Bacon: The Philosophy of Science in the Middle Ages* (Simpkin, Marshall & Co., 1876), 5.

1.Jules Michelet, *Histoire de France*, 19 vols. (Hachette, 1831–67); Jacob Burckhardt, *Die Kultur der Renaissance in Italien* (Schweighauser'schen Verlagsbuchhandlung, 1860), translated by S. G. C. Middlemore as *The Civilization of the Renaissance in Italy* (C. K. Paul & Co., 1878).

2. Émile Charles, *Roger Bacon: Sa vie, ses ouvrages, ses doctrines* (Hachette, 1861), 102. Jules Balteau, ed., *Dictionnaire de biographie français* (Letouzey et Ané, 1932), q.v. "Emile Charles."

3. By this point, I hope you are putting this phrase and the accompanying adjective "medieval" into quotation marks every time you encounter them.

4. This is the period known as the Western Schism (1378–1417); see Joëlle Rollo-Koster and Thomas M. Izbicki, eds., *A Companion to the Great Western Schism (1378–1417)* (Brill, 2009). In the Orthodox Church St. Christopher was known as a member of the Cynocephali, the dog-headed giant people; Saint Guinefort, the Holy Greyhound, was venerated as a saint in thirteenth-century France; see Jean-Claude Schmitt, *The Holy Greyhound: Guinefort, Healer of Children Since the Thirteenth Century*, trans. Martin Thom (Cambridge University Press, 1983). Before the clerical reforms of the eleventh and twelfth centuries, clerical celibacy was not mandatory in the Catholic Church. See, for example, Jennifer D. Thibodeaux, *The Manly Priest: Clerical Celibacy, Masculinity, and Reform in England and Normandy, 1066–1300* (University of Pennsylvania Press, 2015). Abelard claimed that the monks at St. Gilda, in Brittany, tried to poison him; see Peter Abelard, *Historia calamitatum*, in *The Letters of Abelard and Heloise*, ed. and trans. Betty Radice, revised ed. (Penguin, 2004). Secular and spiritual power often had a contentious relationship, such as during the period from the late- eleventh century through the twelfth known as the Investiture Contest; see Brian Tierney, *The Crisis of Church and State: 1050–1300*, Medieval Academy Reprints for Teaching 21 (University of Toronto Press, 1988). For just one necromancer pope, see E. R. Truitt, "Celestial Divination and Arabic Science in Twelfth-Century England: The History of Gerbert of Aurillac's Talking Head," *Journal for the History of Ideas* 73 (2012): 201–22.

5. In the early medieval period, the Irish Catholic Church adhered to different monastic customs and calculated the date of Easter differently from the Roman Catholic Church, leading to disagreements and strife, as well as opposing claims to liturgical legitimacy. Bede writes about this conflict in his *Ecclesiastical History of the English People* (731). On the Rhineland massacres, see Robert Chazan, *In the Year 1096: The First Crusade and the Jews* (Jewish Publication Society, 1996).

6. Adamson, *Roger Bacon*, 3.

7. Adamson, *Roger Bacon*, 5, 19.

8. Adamson, *Roger Bacon*, 16–17.

9. Adamson briefly noted that Jebb's version of *Opus maius* was incomplete; see Adamson, *Roger Bacon*, 21. On this point, see Amanda Power, "A Mirror for Every Age: The Reputation of Roger Bacon," *English Historical Review* 121, no. 492 (2006): 657–92, 677n108.

10. Adamson, *Roger Bacon*, 14.

11. Adamson, *Roger Bacon*, 13.

12. Leslie Stephen, ed., *Dictionary of National Biography* (Smith, Elder, & Co., 1885), q.v. "Roger Bacon," 1:847a.

13. John William Draper, *History of the Conflict Between Religion and Science* (D. Appleton and Co., 1874); Andrew Dickson White, *A History of the Warfare of Science with Theology in Christendom*, 2 vols. (D. Appleton and Co., 1896). See also David C. Lindberg and Ronald L. Numbers, "Beyond War and Peace: A Reappraisal of the Encounter Between Christianity and Science," *Perspectives on Science and Christian Faith* 39, no. 3 (1987): 140–49; James C. Ungureanu, *Science, Religion, and the Protestant Tradition: Retracing the Origins of Conflict* (University of Pittsburgh Press, 2019).

14. White, *A History of the Warfare*, 1:387. White promulgated a number of myths about science in the medieval period. See Katharine Park, "Myth: That the Medieval Church Prohibited Human Dissection," in *Galileo Goes to Jail, and Other Myths About Science and Religion*, ed. Ronald L. Numbers (Harvard University Press, 2009), 43–49. See also the "Introduction" to this volume.

15. Power, "A Mirror for Every Age," 679.

16. Georg Wilhelm Friedrich Hegel, *Lectures on the History of Philosophy*, trans. E. S. Haldane and F. H. Simson, 3 vols. (London, 1892–96), 3:92.

17. The ultimate nightmare!

18. Lorraine Daston, "The History of Science and the History of Knowledge," *KNOW: A Journal on the Formation of Knowledge* 1, no. 1 (2017): 131–54, https://doi.org/10.1086/691678.

19. Power, "A Mirror for Every Age," 676.

20. See Roger Bacon, *Fr. Rogeri Bacon opera quaedam hactenus inedita*, ed. J. S. Brewer (Longman and Green, 1859).

21. Falconer Madan, "Some Interesting Features in the Past History of Science in Oxford," collected in G. A. Oxon., C. 130, vol. 1, fols. 9–38, Oxford, Bodleian Library, quote on fol. 9. This and one other volume contain collected ephemera relating to the commemoration of Roger Bacon at the University of Oxford, 1893–1915. I first learned of these volumes in Power's "A Mirror for Every Age."

22. Madan, "Some Interesting Features," in G. A. Oxon., C. 130, vol. 1, fol. 12; G. C. Bourne, "Mr. Falconer Madan on the Past History of Science at Oxford," *Oxford Magazine*, November 29, 1893, 111–12.

23. Madan, "Some Interesting Features," in G. A. Oxon., C. 130, vol. 1, fol. 15.

24. Bourne, "Mr. Falconer Madan," 112.

25. My thanks to John Landsdowne for bringing the science pillar at St. Anne's Cathedral to my attention.

26. G. A. Oxon., C. 130, vol. 1, fol. 66 (undated), Bodleian Library, Oxford University.

27. G. A. Oxon., C. 130, vol. 1, fol. 68, Bodleian Library, Oxford University.

28. Pierre Duhem, *Sozein ta phainomena, essai sur la notion de théorie physique de Platon à Galilée* (Hermann, 1908), translated into English by Edmond Dolland and Chaninah Maschler as *To Save the Phenomena* (University of Chicago Press, 1969); and Pierre Duhem, *Le Système du monde: Histoire des doctrines cosmologiques de Platon à Copernic*, 10 vols. (Hermann, 1913–1959); Ludwig Baur, *Die philosophischen Werke des Robert Grosseteste, Bischofs von Lincoln*, Beiträge zur Geschichte der Philosophie und Theologie des Mittelalters 9 (Aschendorff, 1912).

29. Miers, a chemist, was the principal (head) of the University of London; Lucien Poincaré, a physicist, (brother of Henri Poincaré and brother-in-law to Émile Boutroux, mentioned above) was the director of secondary and then higher education at the Ministry of Public Instruction in France.

30. Cloudesley Shovell Henry Brereton and Garnet Smith, "Roger Bacon Once More," *Times Literary Supplement*, June 25, 1914, 309. Brereton, who was one of Bergson's translators into English, was also a member of the General Committee. See also John Edwin Sandys, "Roger Bacon," *Saturday Review of Politics, Literature, Science, and Art*, June 1914, appearing also in *Homiletic Review: An International Magazine of Religion, Theology, and Philosophy*,

June 1914, 433–38; "The Roger Bacon Commemoration at Oxford," *Nature* 93, no. 2327 (June 4, 1914): 355–55 (published in advance of the event); "The Commemoration of Roger Bacon at Oxford," *Nature* 93, no. 2329 (June 18, 1914): 405–6 (a brief account of the event); "Roger Bacon," *British Medical Journal* (June 20, 1914): 1365–367.

31. Cecil Underwood, "Roger Bacon," *Ave Maria*, June 20, 1914, 769–75; D. O'Keefe, "Roger Bacon," *Irish Ecclesiastical Record*, September 1914, 346–357.

32. *The Open Court: Devoted to the Science of Religion, the Religion of Science, and the Extension of the Religious Parliament Idea*, August 1918; A. G. Little, ed., *Roger Bacon: Essays Contributed by Various Writers on the Occasion of the Commemoration of the Seventh Centenary of his Birth* (Clarendon Press, 1914). On the impact of *The Open Court* and its editor, Paul Carus, in bringing science and religion together in the public sphere, see Constance Myers, "Paul Carus and *The Open Court*: History of a Journal," *Midcontinent American Studies Journal* 5, no. 2 (1964): 57–68.

33. John Erskine, *A Pageant of the Thirteenth Century for the Seven Hundredth Anniversary of Roger Bacon*, plan by John J. Coss (Columbia University Press, 1914). Although the play was never performed Erskine maintained a vigilant interest in his copyright; see Columbia University Press Records, Box CC4, Rare Book and Manuscript Library, Columbia University Libraries.

34. Erskine, *A Pageant of the Thirteenth Century*, 35.

35. Erskine, *A Pageant of the Thirteenth Century*, 8.

36. Under the early 'Abbasid caliph Al-Mansur (r. 754–75 C.E. / 136–58 A.H.), its primary function was the translation of Sasanian Persian historical and cultural texts into Arabic; under Al-Ma'mūn (813–33 C.E. / 196–216 A.H.) the bureau appears to have been tasked also with translating material related to astral science and mathematics. See Dimitri Gutas, *Greek Thought, Arabic Culture: The Graeco-Arabic Translation Movement in Baghdad and Early 'Abbāsid Society (2nd–4th/8th–10th Centuries)* (Routledge, 1998), 54–60.

37. Erskine, *A Pageant of the Thirteenth Century*, 26, 38.

38. Ungureanu, *Science, Religion, and the Protestant Tradition*, 250–55; Power, "A Mirror for Ever Age," 677–79. On Sarton's view of the history of science as a metascientific discipline, see Peter Dear, "The History of Science and the History of the Sciences: George Sarton, *Isis*, and the Two Cultures," *Isis* 100, no. 1 (2009): 89–93.

39. Gilbert Allardyce, "The Rise and Fall of the Western Civilization Course," *American Historical Review* 87, no. 3 (1982): 695–725.

40. Louis Menand, *The Marketplace of Ideas: Reform and Resistance in the American University* (W. W. Norton, 2010), 35. On the term "Anglo-Saxon," see Adam Miyashiro, "Our Deeper Past: Race, Settler Colonialism, and Medieval Heritage Politics," *Literature Compass* 16, no. 9–10 (2019): https://doi.org/10.1111/lic3.12550; Matthew X. Vernon, *The Black Middle Ages: Race and the Construction of the Middle Ages* (Palgrave Macmillan, 2018).

41. For example, Herbert Hochberg, "The Empirical Philosophy of Roger and Francis Bacon," *Philosophy of Science* 20, no. 4 (1953): 313–26, 314.

42. Duhem, *Le système du monde*, 3:442; Lynn Thorndike Jr., "Roger Bacon and Experimental Method in the Middle Ages," *Philosophical Review* 23, no. 3 (1914): 271–98.

43. Lynn Thorndike Jr., *A History of Magic and Experimental Science*, 8 vols. (Columbia University Press, 1923–58), 2:618.

44. George Sarton, *Introduction to the History of Science*, 3 vols. (Johns Hopkins University Press, 1931), 2:953.

45. Sarton, *Introduction to the History of Science*, 2:954.

46. Duhem, *Le système du monde.*

47. Alexandre Koyré, *Études Galiléennes*, 3 vols. (Hermann, 1939), 1:9; Alexandre Koyré, "Galileo and Plato," *Journal of the History of Ideas* 4, no. 4 (1943): 400–428.

48. Alistair Crombie, *Robert Grosseteste and the Origins of Experimental Science, 1100–1700*, 2 vols. (Clarendon Press, 1953); Alistair Crombie, "The Significance of Medieval Discussions of Scientific Method for the Scientific Revolution," in *Critical Problems in the History of Science*, ed. Marshall Clagett (University of Wisconsin Press, 1959), 79–101. John Murdoch, just a few years later, argued in favor of Koyré's position. John Murdoch, "Philosophy and the Enterprise of Science in the Later Middle Ages," in *The Interaction Between Science and Philosophy*, ed. Yehuda Elkana and Samuel Sambursky (Humanities Press, 1974), 51–74; John Murdoch, "Pierre Duhem and the History of Late Medieval Science and Philosophy in the Latin West," in *Gli studi di filosofia medievale fra Otto e Novecento*, ed. Imbach Ruedi and Maierù Alfonso (Edizioni di Storia e Letteratura, 1991), 253–302.

49. David C. Lindberg, *The Beginnings of Western Science: The European Scientific Tradition in Philosophical, Religious, and Institutional Contexts, Prehistory to A. D. 1450* (University of Chicago Press, 1992), esp. 355–68; Edward Grant, *Physical Science in the Middle Ages* (Wiley, 1971); and Edward Grant, *The Foundations of Modern Science in the Middle Ages* (Cambridge University Press, 1996).

50. Power, "A Mirror for Every Age," 688; Patricia Clare Ingham, *The Medieval New: Ambivalence in an Age of Innovation* (University of Pennsylvania Press, 2015), 50–55.

51. Crombie, *Robert Grosseteste and the Origins of Experimental Science*, 1:156.

52. Stewart C. Easton, *Roger Bacon and His Search for a Universal Science: A Reconsideration of the Life and Work of Roger Bacon in the Light of His Own Stated Purposes* (Columbia University Press, 1952); Theodore Crowley, *Roger Bacon: The Problem of the Soul in His Philosophical Commentaries* (Duffy, 1950). See also Power, "A Mirror for Every Age," 690.

53. Thomas S. Kuhn, *The Structure of Scientific Revolutions*, 4th ed. (University of Chicago Press, 2012), 167. On the similar narratives of progress and regimes of historicity that underpin both narratives of artistic and scientific development, see Gianni Vattimo, *La fine della modernità* (Garzanti Editore, 1985), in English as *The End of Modernity: Nihilism and Hermeneutics in Postmodern Culture*, ed. and trans. Jon R. Snyder (Johns Hopkins University Press, 1988), 90–93.

54. Despite their invention as tools to help achieve these goals, the *MGH* and others remain in wide use by scholars and students around the world as they often contain the only or most complete printed edition of a text or group of texts. On the role of medieval scholarship (and scholars) in nineteenth and early-twentieth century nationalism, see R. Howard Bloch and Stephen G. Nichols, eds., *Medievalism and the Modernist Temper* (Johns Hopkins University Press, 1996). On the role of medieval scholarship to support nationalism at the start of the twenty-first century, see Bruce Holsinger, *Neomedievalism, Neoconservatism, and the War on Terror* (University of Chicago Press, 2007).

55. Lindberg, *The Beginnings of Western Science*; Guy Attewell, "Islamic Medicines: Perspectives on the Greek Legacy in the History of Islamic Medical Traditions in West Asia," in *Medicine Across Cultures: History and Practice of Medicine in Non-Western Cultures*, ed. Helaine Selin (Springer, 2003), 325–50; Toby E. Huff, *The Rise of Early Modern Science: Islam, China, and the West*, 2nd ed. (Cambridge University Press, 2003); H. Floris Cohen, *How Modern Science Came into the World: Four Civilizations, One 17th-Century Breakthrough*

(Amsterdam University Press, 2012); David Wootton, *The Invention of Science: A New History of the Scientific Revolution* (HarperCollins, 2015). On how the historiography of Islamic science in this period has remained tethered to ideas about the centrality of Greek philosophy and the "failure" to develop into experimental science, see Ahmed Ragab, "Translation and the Making of a Medical Archive: The Case of the Islamic Translation Movement," *Osiris* 37, no. 1 (2022): 25–46; George Saliba, *Islamic Science and the Making of the European Renaissance* (MIT Press, 2007). Bruno Latour argued that the term "modernity" designates the winners and losers in a fight. Bruno Latour, *Nous n'avons jamais été modernes: Essai d'anthropologie symétrique* (La Decouverte, 1991), in English as *We Have Never Been Modern*, trans. Catherine Porter (Harvard University Press, 1993), 10.

CONCLUSION

Note to epigraph. Eugene Hecht, *Optics*, 5th ed. (Pearson, 2016), 2. This is also erroneous; Bacon influenced Witelo, not the other way around.

1.George Sarton, *Introduction to the History of Science*, 3 vols. (Johns Hopkins University Press, 1931), 2:954.

2. See Monica Green, *The Trotula: A Medieval Compendium of Women's Medicine* (University of Pennsylvania Press, 2003); Paula Findlen, "Inventing the Middle Ages: An Early Modern Forger Hiding in Plain Sight," in *For the Sake of Learning: Essays in Honor of Anthony Grafton*, ed. Ann Blair and Anja-Silvia Goering, 2 vols. (Brill, 2016), 2:871–96; Tara Nummedal, *Anna Zieglerin and the Lion's Blood: Alchemy and End Times in Reformation Germany* (University of Pennsylvania Press, 2019); and Leah DeVun, *The Shape of Sex: Nonbinary Gender from Genesis to the Renaissance* (Columbia University Press, 2021).

3. See "Introduction."

4. David C. Lindberg and Ronald L. Numbers, "Beyond War and Peace: A Reappraisal of the Encounter Between Christianity and Science," *Perspectives on Science and Christian Faith* 39, no. 3 (1987): 140–49; James C. Ungureanu, *Science, Religion, and the Protestant Tradition: Retracing the Origins of Conflict* (University of Pittsburgh Press, 2019); Steven Shapin, *The Scientific Revolution*, 2nd ed. (University of Chicago Press, 2018).

5. Toby E. Huff, *Intellectual Curiosity and the Scientific Revolution: A Global Perspective* (Cambridge University Press, 2010); James Poskett, *Horizons: The Global Origins of Modern Science* (Mariner Books, 2022).

6. Pablo F. Gómez, *The Experiential Caribbean: Creating Knowledge and Healing in the Early Modern Atlantic* (University of North Carolina Press, 2017); Pamela H. Smith, *From Lived Experience to the Written Word: Reconstructing Practical Knowledge in the Early Modern World* (University of Chicago Press, 2022); Elaine Leong, *Recipes and Everyday Knowledge: Medicine, Science, and the Household in Early Modern England* (University of Chicago Press, 2018); Alisha Rankin, *The Poison Trials: Wonder Drugs, Experiment, and the Battle for Authority in Renaissance Science* (University of Chicago Press, 2021).

7. Amanda Power, "A Mirror for Every Age: The Reputation of Roger Bacon," *English Historical Review* 121, no. 492 (2006): 657–92, 676.

8. Hecht, *Optics*, 1.

9. Jeffrey O. Bennett, Megan O. Donahue, Nicholas Schneider, and Mark Voit, *The Cosmic Perspective*, 9th ed. (Pearson, 2020), 60–63; Paul A. Tipler and Ralph A. Llewellyn, *Modern Physics*, 6th ed. (W. H. Freeman and Co., 2012); Douglas C. Giancoli, *Physics: Principles*

with Applications, 7th ed. (Pearson, 2014); Barbara Ryden and Bradley M. Peterson, *Foundations of Astrophysics* (Addison-Wesley, 2010).

10. Ryden and Peterson, *Foundations of Astrophysics*, 1.

11. Umberto Eco, *The Name of the Rose*, trans. William Weaver (Harcourt, 1983); George R. R. Martin, *A Game of Thrones* (Bantam, 1996).

12. For example, see Joanna Kavenna, *A Field Guide to Reality* (Riverrun, 2017); Peter Darvill-Evans, *Asylum (Dr. Who)* (BBC Books, 2001); James A. Owens, *Here, There Be Dragons* (Simon and Schuster, 2010); Rachel Cantor, *A Highly Unlikely Scenario* (Melville House, 2014).

13. David Graeber and David Wengrow, *The Dawn of Everything: A New History of Humanity* (Farrar, Straus and Giroux, 2021).

Bibliography

UNPUBLISHED SOURCES

Bacon, Roger. *Commentary on The Secret of Secrets.* MS Tanner 116. Oxford, Bodleian Library.

———. *Opus Maius.* MS Digby 235. Oxford, Bodleian Library.

———. *Perspectiva*, in *Opus Maius.* MS Savile 18. Oxford, Bodleian Library.

Bestiary. Harley MS 4751. London, British Library.

Bible moralisée. MS Harley 1527. London, British Library.

de Morgan, Augustus. Letter to John Herschel, February 11, 1859. HS/6/324. London, The Royal Society.

Hooke, Robert. "Lecture on the 'arithmatick of the Brachmans' and of a sort of 'Carolina cloath.'" 1689. CLP/20/79. London, The Royal Society.

Landseer, Edwin. *Queen Victoria and Prince Albert at the Bal Costumé of 12 May 1842.* 1842–1846. RCIN 404540. London, Royal Collection Trust.

Le Livre et le vraye hystoire du bon roy Alixandre. Royal MS 20 B XX. London, British Library.

L'Ystoire du bon roi Alexandre. 1320. MS 78 C 1. Berlin, Kupferstichkabinett, Staatliche Museum Preußischer Kulturbesitz.

Madan, Falconer. "Some Interesting Features in the Past History of Science in Oxford." Collected in G. A. Oxon. C. 130, 2 vols. Oxford, Bodleian Library.

Peter of Trau. *Chronicle.* MS Canon. Misc. 525. Oxford, Bodleian Library.

Pseudo-Aristotle. *Kitāb Sirr al-asrār.* LJS 459. Philadelphia, Library of the University of Pennsylvania.

Pseudo-Aristotle. *Secretum Secretorum.* 1326. MS Add. 47680. London, British Library.

Whewell, William. Research Notebook. n.d. R. 18.10 (1). Cambridge, Trinity College Library.

———. Research Notebook. n.d. R. 18.17 (13). Cambridge, Trinity College Library.

———. Research Notebook. n.d. R. 18.17 (15). Cambridge, Trinity College Library.

PUBLISHED SOURCES

Abelard, Peter. *Historia calamitatum.* In *The Letters of Abelard and Heloise*, edited and translated by Betty Radice. Revised ed. Penguin, 2004.

Abū Maʿšar. *Abū Maʿšar on Historical Astrology: The Book of Religions and Dynasties (On the Great Conjunctions)*. Edited by Keiji Yamamoto and Charles Burnett. 2 vols. Brill, 2000.

Adamson, Peter. *Al-Kindī*. Oxford University Press, 2006.

Adamson, Peter, and Richard C. Taylor. "Introduction." In *Cambridge Companion to Arabic Philosophy*, edited by Peter Adamson and Richard Taylor. Cambridge University Press, 2005.

Adamson, Robert. *Roger Bacon: The Philosophy of Science in the Middle Ages*. Simpkin, Marshall & Co., 1876.

Agrimi, Jole, and Chiara Crisciani. "Albumazar nell'astrologia di Ruggero Bacone." *Acme* 25, no. 3 (1972): 315–38.

Ahmed, Shahab. *What Is Islam? The Importance of Being Islamic*. Princeton University Press, 2015.

Akbari, Suzanne Conklin. *Seeing Through the Veil*. University of Toronto Press, 2004.

Alessio, Franco. *Mito e scienza in Ruggero Bacone*. Ceschina, 1957.

Alexandre de Paris. *Le roman d'Alexandre*. Translated by Laurence Harf-Lancner. Livre de poche, 1994.

———. *Le roman d'Alexandre ou le roman de toute chevalerie*. Translated by Catherine Gaullier-Bougassas and Laurence Harf-Lancner. Honoré Champion, 2003.

———. *The Medieval French Roman d'Alexandre*. Vol. 2, *Version of Alexandre de Paris*. Edited by Edward C. Armstrong, Douglas L. Buffum, Bateman Edwards, and L. F. H. Lowe. Princeton University Press, 1937.

Al-Kindī. *De radiis*. Edited by Marie-Thérèse d'Alverny and Françoise Hudry. Librairie J. Vrin, 1975.

Allardyce, Gilbert. "The Rise and Fall of the Western Civilization Course." *American Historical Review* 87, no. 3 (1982): 695–725.

Allen, Meagan S. "Alchemy and Pharmacology in the *Opus maius*." In *The Sciences of Roger Bacon's* Opus maius, edited by Meagan S. Allen. Routledge, 2025.

———. *Roger Bacon and the Incorruptible Human, 1220–1292: Alchemy, Pharmacology, and the Desire to Prolong Life*. Palgrave Studies in Medieval and Early Modern Medicine. Palgrave Macmillan, 2023.

———. "Roger Bacon's Medical Alchemy and the Multiplication of Species." In *The Philosophy and Science of Roger Bacon: Studies in Honour of Jeremiah Hackett*, edited by Nicola Polloni and Yael Kedar. Routledge, 2021.

———, ed. *The Sciences of Roger Bacon's* Opus maius. Routledge, 2025.

Allen, Michael I. "Universal History 300–1000: Origins and Western Developments." In *Historiography in the Middle Ages*, edited by Deborah Mauskopf Deliyannis. Brill, 2006.

Altschul, Nadia R. "Postcolonialism and the Study of the Middle Ages." *History Compass* 6, no. 2 (2008): 588–606.

Analecta Franciscana. Vol. 3, *Chronica XXIV generalium Ordinis minorum*. Quaracchi, 1897.

Anheim, Étienne, Benoît Grévin, and Martin Morard. "Exégèse judéo-chrétienne, magie, et linguistigue: Un recueil de *Notes* inédites attribuées à Roger Bacon." *Archives d'Histoire Doctrinale et Littéraire du Moyen Âge* 68 (2001): 95–154.

Aristotle. *Metaphysics*. Vol. 1, *Books 1–9*. Translated by Hugh Tredennick. Loeb Classical Library 271. Harvard University Press, 1933.

Arnold, John H. *What Is Medieval History?* Polity, 2008.

Ashworth, William J. *The Trinity Circle: Anxiety, Intelligence, and Knowledge Creation in Nineteenth-Century England.* University of Pittsburgh Press, 2021.

Attewell, Guy. "Islamic Medicines: Perspectives on the Greek Legacy in the History of Islamic Medical Traditions in West Asia." In *Medicine Across Cultures: History and Practice of Medicine in Non-Western Cultures*, edited by Helaine Selin. Springer, 2003.

Aubrey, John. *Brief Lives.* Edited by Andrew Clark. Oxford: Clarendon, 1898.

———. *Brief Lives with an Apparatus for the Lives of our English Mathematical Writers.* Edited by Kate Bennett. Oxford University Press, 2015.

Auerbach, Erich. "Figura." In *Scenes from the Drama of European Literature.* University of Minnesota Press, 1994.

Bacon, Francis. *Novum organum.* Edited by Thomas Fowler. Clarendon Press, 1878 [John Bill, 1620].

———. "Temporis partus masculus." In *The Works of Francis Bacon*, vol. 3, *Philosophical Works*, edited by James Spedding, Robert Leslie Ellis, and Douglas Denon Heath. Reprint, Garret Press, 1968 [Longmans, 1857].

Bacon, Frier. *His Discovery of the Miracles of Art, Nature, and Magick, Faithfully Translated out of Dr. Dee's Own Copy by T. M., and Never Before in English.* London: Printed for Simon Miller, 1659.

Bacon, Roger. *Communia naturalium.* Edited by Robert Steele. Clarendon Press, 1909.

———. *Compendium of the Study of Philosophy.* Edited and translated by Thomas S. Maloney. Oxford University Press, 2018.

———. *Cure of Old Age and the Preservation of Youth by Roger Bacon, Translated out of Latin, with Annotations, and an Account of His Life and Writings.* Edited by Richard Browne. Printed for Tho. Flesher . . . and Edward Evets, 1683.

———. "Epistola Fratris Rogeri Baconi." In *Lettera a Clemente IV*, edited and translated by P. Efrem Bettoni. Biblioteca Francescana Provinciale, 1964.

———. *Epistolae fratris Rogerii Baconis, De secretis operibus artis et naturae, et de nullitate magiae, opera Iohannis Dee Londinensis e pluribus exemplaribus castigata olim, et ad sensum integrum restituta: Nunc vero a quodam veritatis amatore, in gratiam verae scientiae candidatorum foras emissa, cum notis quibusdam partim ipsius Johannis Dee, partim edentis.* Ex Bibliopolio Frobeniano, 1618.

———. *Fr. Rogeri Bacon opera quædam hactenus inedita.* Edited by J. S. Brewer. Longman and Green, 1859.

———. *Moralis philosophia.* Edited by Eugenio Massa. In Ædibus Thesauri mundi, 1953.

———. *Opera hactenus inedita Rogeri Baconi.* Fasc. 5, *Secretum secretorum cum glossis et notulis: Tractatus brevis et utilis ad declarandum quedam obscure dicta Fratris Rogeri.* Edited by Robert Steele. Clarendon Press, 1920.

———. *Opera hactenus inedita Rogeri Baconi.* Fascs. 7, 8, 10–15. *Quaestiones.* Edited by Robert Steele. Clarendon Press, 1920.

———. *Opus majus.* Translated by Robert Belle Burke. 2 vols. University of Pennsylvania Press, 1928.

———. *Opus tertium.* Edited and translated by Nikolaus Egel. Felix Meiner, 2020.

———. *Opus majus.* Edited by John Henry Bridges. 3 vols. Clarendon Press, 1897–1900.

———. *Opus majus.* Edited by Samuel Jebb. William Bowyer, 1733.

———. *Part of the Opus Tertium of Roger Bacon, Including a Fragment Now Printed for the First Time.* Edited by A. G. Little. Aberdeen University Press, 1912.

———. *Un fragment inédit de l'Opus tertium de Roger Bacon précédé d'une étude sur ce fragment.* Edited by Pierre Duhem. Quarrachi, 1909.

Bacone, Ruggero [Roger Bacon]. *Filosofia, scienza, teologia dall'Opus Maius.* Edited and translated by Valeria Sorge and Fabio Seller. Armando, 2010.

——— [Roger Bacon]. *La scienza sperimentale: Lettera a Clemente IV—La scienza sperimentale—I segreti dell'arte e della natura.* Translated by Francesco Bottin. Rusconi, 1990.

Baldwin, John W. *The Scholastic Culture of the Middle Ages, 1000–1300.* Wayland, 1997 [Heath, 1971].

Bale, John. *Scriptorum illustrium maioris Britanniae . . . Catalogus.* 2nd ed. Basel, 1559 [John Overton, 1548].

Balteau, Jules, ed. *Dictionnaire de biographie française.* Letouzey et Ané, 1932.

Bataillon, Louis-Jacques. "Guillaume de la Mare: Note sur sa régence parisienne et sa prédication." *Archivum Franciscanum Historicum* 98 (2005): 367–422.

Baur, Ludwig. *Die philosophischen Werke des Robert Grosseteste, Bischofs von Lincoln.* Beiträge zur Geschichte der Philosophie und Theologie des Mittelalters 9. Aschendorff, 1912.

Becher, Harvey W. "William Whewell's Odyssey: From Mathematics to Moral Philosophy." In *William Whewell: A Composite Portrait*, edited by Menachem Fisch and Simon Schaffer. Oxford University Press, 1991.

Beckwith, Christopher I. *Empires of the Silk Road: A History of Central Eurasia from the Bronze Age to the Present.* Princeton University Press, 2009.

Belting, Hans. *Baghdad and Florence: Renaissance Art and Arab Science.* Translated by Deborah Lucas Schneider. Harvard University Press, 2011.

Bénatouïl, Thomas, and Isabelle Draelants, eds. *Expertus sum: L'expérience par les sens dans la philosophie naturelle médiévale.* Micrologus 40. SISMEL Edizioni del Galluzzo, 2011.

Bennett, Jeffrey O., Megan O. Donahue, Nicholas Schneider, and Mark Voit. *The Cosmic Perspective.* 9th ed. Pearson, 2020.

Bennett, Judith M. *History Matters: Patriarchy and the Challenge of Feminism.* University of Pennsylvania Press, 2006.

Bergdolt, Klaus "Bacon und Giotto: Zum Einfluß der franziskanischen Naturphilosophie auf die bildende Kunst am Ende des 13. Jahrhunderts." *Medizinhistorisches Journal* 24 (1989): 25–41.

Bernardini, Paola, and Anna Rodolfi, eds. *Roger Bacon's* Communia Naturalium*: A 13th Century Philosopher's Workshop.* Micrologus 64. SISMEL Edizioni del Galluzzo, 2014.

Bérubé, Camille. *De la philosophie à la sagesse chez Saint Bonaventure et Roger Bacon.* Istituto Storico dei Cappuccini, 1976.

Bevilacqua, Alexander. *The Republic of Arabic Letters: Islam and the European Enlightenment.* Belknap Press of Harvard University Press, 2018.

Biddick, Kathleen. *The Typological Imaginary: Circumcision, Technology, History.* University of Pennsylvania Press, 2003.

Bigalli, Davide. *I Tartari e l'Apocalisse: Ricerche sull'escatologia in Adamo Marsh e Ruggero Bacone.* La Nuova Italia, 1971.

Bihl, Michael. "Statuta generalia ordinis edita in capitulis generalibus celebratis Narbonae an. 1260, Assisii an. 1279 atque Parisiis an. 1292." *Archivum Franciscanum Historicum* 34, no. 1 (1941): 13–94.

Birch, Thomas. "July, 1682." In *The History of the Royal Society of London for Improving of Natural Knowledge, from Its First Rise in Which the Most Considerable of Those Papers*

Communicated to the Society, Which Have Hitherto Not Been Published, Are Inserted in Their Proper Order, as a Supplement to the Philosophical Transactions. Vol. 4. A. Millar, 1757.

———. "March/April 1679." In *The History of the Royal Society of London for Improving of Natural Knowledge, from Its First Rise in Which the Most Considerable of Those Papers Communicated to the Society, Which Have Hitherto Not Been Published, Are Inserted in Their Proper Order, as a Supplement to the Philosophical Transactions.* Vol. 3. A. Millar, 1757.

Bloch, R. Howard, and Stephen G. Nichols, eds. *Medievalism and the Modernist Temper.* Johns Hopkins University Press, 1996.

Blumenberg, Hans. *The Legitimacy of the Modern Age.* Translated by Robert M. Wallace. MIT Press, 1983.

Boer, W. Walther, ed. *Epistola Alexandri Macedonis ad Aristotelem magistrum suum de itinere suo et de situ Indiae.* Beiträge zur klassischen Philologie. Hain, 1980.

Boggs, Abigail, Eli Meyerhoff, Nick Mitchell, and Zach Schwartz-Weinstein. "Abolitionist University Studies: An Invitation." *Abolition University* (blog), accessed August 25, 2024. https://abolition.university/wp-content/uploads/2019/08/Abolitionist-University-Studies_-An-Invitation-Release-1-version.pdf.

Boggs, Abigail, and Nick Mitchell. "Critical University Studies and the Crisis Consensus." *Feminist Studies* 44, no. 2 (2018): 432–63.

Bourne, G. C. "Mr. Falconer Madan on the Past History of Science at Oxford." *Oxford Magazine*, November 29, 1893, 111–12.

Brereton, Cloudesley Shovell Henry, and Garnet Smith. "Roger Bacon Once More." *Times Literary Supplement*, June 25, 1914, 309.

Brown, Catherine. "In the Middle." *Journal of Medieval and Early Modern Studies* 30, no. 3 (2000): 547–74.

Bruner, Otto, Werner Conze, and Reinhart Koselleck, eds. Introduction. *Geschichtliche Grundbegriffe: Historisches Lexikon zur politisch-sozialen Sprache in Deutschland.* Klett-Cotta, 1972.

Buellens, Pieter. "Robert Grosseteste and the Fluid History of the Latin *Nicomachean Ethics.*" *Revista Española de Filosofía Medieval* 30, no. 1 (2023): 177–98.

Burckhardt, Jacob. *The Civilization of the Renaissance in Italy.* Translated by S. G. C. Middlemore. C. K. Paul & Co., 1878.

———. *Die Kultur der Renaissance in Italien.* Schweighauser'schen Verlagsbuchhandlung, 1860.

Burnett, Charles. *The Introduction of Arabic Learning into England.* The British Library, 1997.

———. "Michael Scot and the Transmission of Scientific Culture from Toledo to Bologna via the Court of Frederick II Hohenstaufen." In *Le scienze alla corte di Federico II.* Micrologus 2. SISMEL Edizioni del Galluzzo, 1994.

Caciola, Nancy Mandeville. *Discerning Spirits: Divine and Demonic Possession in the Middle Ages.* Cornell University Press, 2003.

Cadden, Joan. *The Meanings of Sex Difference in the Middle Ages: Medicine, Science, and Culture.* Cambridge University Press, 1993.

———. *Nothing Natural Is Shameful: Sodomy and Science in Late Medieval Europe.* University of Pennsylvania Press, 2013.

Calendar of the Patent Rolls Preserved in the Public Record Office, Henry III (1258–1266). Public Record Office, 1910.

Campbell, Mary Baine. *Witness and the Other World: Exotic European Travel Writing, 400–1600*. Cornell University Press, 1988.

Cantor, Geoffrey N. "Between Rationalism and Romanticism: Whewell's Historiography of the Inductive Sciences." In *William Whewell: A Composite Portrait*, edited by Menachem Fisch and Simon Schaffer. Oxford University Press, 1991.

Cantor, Rachel. *A Highly Unlikely Scenario*. Melville House, 2014.

Carey, Hilary M. "Astrology and Antichrist in the Later Middle Ages." In *Time and Eternity: The Medieval Discourse*, edited by Gerhard Jaritz and Gerson Moreno-Riaño. Brepols, 2003.

Carpenter, David. *Henry III: The Rise to Power and Personal Rule, 1207–1258*. Yale University Press, 2020.

Carton, Raoul. *L'expérience physique chez Roger Bacon*. J. Vrin, 1924.

Cary, George. *The Medieval Alexander*. Edited by David J. A. Ross. Cambridge University Press, 1956.

Cellarius, Christoph. *Historia universalis breviter ac perspicue exposita, in antiquam, et medii aevi ac novam divisa, cum notis perpetuis*. Jena, 1688.

Chakrabarty, Dipesh. "Historicism and Its Supplements: A Note on a Predicament Shared by Medieval and Postcolonial Studies." In *Medievalisms in the Postcolonial World: The Idea of "the Middle Ages" Outside Europe*, edited by Kathleen Davis and Nadia Altschul. Johns Hopkins University Press, 2009.

———. "The Muddle of Modernity." *American Historical Review* 116, no. 3 (2011): 663–75.

———. "Postcoloniality and the Artifice of History: Who Speaks for 'Indian' Pasts?" *Representations* 37 (1992): 1–26.

———. *Provincializing Europe: Postcolonial Thought and Historical Difference*. Princeton University Press, 2007 [2000].

Charles, Émile. *Roger Bacon: Sa vie, ses ouvrages, ses doctrines*. Hachette, 1861.

Chazan, Robert. *In the Year 1096: The First Crusade and the Jews*. Jewish Publication Society, 1996.

Clanchy, Michael T. "*Moderni* in Education and Government in England." *Speculum* 50, no. 4 (1975): 671–88.

Clark, Stuart. *Vanities of the Eye: Vision in Early Modern European Culture*. Oxford University Press, 2007.

Clegg, Brian. *The First Scientist: A Life of Roger Bacon*. Carroll and Graf, 2003.

Clulee, Nicholas H. "Astrology, Magic, and Optics: Facets of John Dee's Early Natural Philosophy." *Renaissance Quarterly* 30, no. 4 (1977): 632–80.

———. "John Dee's Mathematics and the Grading of Compound Qualities." *Ambix* 18 (1971): 178–211.

Cohen, H. Floris. *How Modern Science Came into the World: Four Civilizations, One 17th-Century Breakthrough*. Amsterdam University Press, 2010.

Cohen, Jeffrey Jerome, ed. *The Postcolonial Middle Ages*. Palgrave, 2000.

Collins, David J. "Albertus, 'Magnus' or Magus? Magic, Natural Philosophy, and Religious Reform in the Late Middle Ages." *Renaissance Quarterly* 63, no. 1 (2010): 1–44.

———. *Disenchanting Albert the Great: The Life and Afterlife of a Medieval Magician*. Penn State University Press, 2024.

"The Commemoration of Roger Bacon at Oxford." *Nature* 93, no. 2329 (June 18, 1914): 405–6.

Conant, Jonathan. *Staying Roman: Conquest and Identity in Africa and the Mediterranean, 439–700*. Cambridge University Press, 2012.

Cooper, Lisa H. "Figures for 'Gretter Knowing:' Forms in the Treatise on the Astrolabe." In *Chaucer and the Subversion of Form*, edited by Thomas A. Prendergast and Jessica Rosenfeld. Cambridge University Press, 2018.

Cowles, Henry M. *The Scientific Method: An Evolution of Thinking from Darwin to Dewey*. Harvard University Press, 2020.

Crisciani, Chiara. "Ruggero Bacone e l' 'Aristotele' del *Secretum secretorum*." In *Christian Readings of Aristotle from the Middle Ages to the Renaissance*, edited by Luca Bianchi. Brepols, 2011.

Crombie, Alistair C. *Robert Grosseteste and the Origins of Experimental Science, 1100–1700*. 2 vols. Clarendon Press, 1953.

———. "The Significance of Medieval Discussions of Scientific Method for the Scientific Revolution." In *Critical Problems in the History of Science*, edited by Marshall Clagett. University of Wisconsin Press, 1959.

Crombie, Alistair C., and J. D. North. "Roger Bacon." In *Dictionary of Scientific Biography*. Charles Scribner's Sons, 1970.

Crowley, Theodore. *Roger Bacon: The Problem of the Soul in His Philosophical Commentaries*. Duffy, 1950.

Curtius, Ernst Robert. *European Literature and the Latin Middle Ages*. Translated by Willard R. Trask. Pantheon Books, 1953.

Dagenais, John, and Margaret Rich Greer. "Decolonizing the Middle Ages: Introduction." In "Decolonizing the Middle Ages." Special issue, *Journal of Medieval and Early Modern Studies* 30, no. 3 (2000): 431–48.

Darvill-Evans, Peter. *Asylum (Dr. Who)*. BBC Books, 2001.

Daston, Lorraine. "The History of Science and the History of Knowledge." *KNOW: A Journal on the Formation of Knowledge* 1, no. 1 (2017): 131–54. https://doi.org/10.1086/691678.

Daston, Lorraine, and Katharine Park. *Wonders and the Order of Nature, 1150–1750*. Zone, 1998.

Davis, Kathleen. *Periodization and Sovereignty: How Ideas of Feudalism and Secularization Govern the Politics of Time*. University of Pennsylvania Press, 2008.

de Fouchécour, Charles-Henri. *Moralia: Les notions morales dans la littérature persane du 3e/9e au 7e/13e siècle*. Paris: Éditions Recherche sur les Civilisations, 1986.

Dear, Peter. "The History of Science and the History of the Sciences: George Sarton, *Isis*, and the Two Cultures," *Isis* 100, no. 1 (2009): 89–93.

———. *Revolutionizing the Sciences: European Knowledge and Its Ambitions, 1500–1700*. Princeton University Press, 2001.

Dee, John. *Mathematicall Praeface to the Elements of Geometrie of Euclid of Megara*. London, 1570.

———. *Propaedeumata aphoristica*. Henricus Suttonus, 1558.

Deliyannis, Deborah Mauskopf. "Introduction." In *Historiography in the Middle Ages*, edited by Deborah Mauskopf Deliyannis. Brill, 2006.

DeVun, Leah. *Prophecy, Alchemy, and the End of Time: John of Rupescissa in the Late Middle Ages*. Columbia University Press, 2009.

———. *The Shape of Sex: Nonbinary Gender from Genesis to the Renaissance*. Columbia University Press, 2021.

Digges, Thomas, and Leonard Digges. *An Arithmeticall Militare Treatise Named Stratioticos . . . Long Since Attempted by Leonard Digges, Gentleman: Augmented, Digested and Lately Finished by Thomas Digges, His Sonne*. Henrie Bynneman, 1579.

Dillon, Sarah M. *Seeing Renaissance Glass: Art, Optics, and Glass of Early Modern Italy, 1250–1425*. Peter Lang, 2018.

Dionisotti, A. C. "On the Greek Studies of Robert Grosseteste." In *The Uses of Greek and Latin: Historical Essays*, edited by A. C. Dionisotti, Anthony Grafton, and Jill Kraye. The Warburg Institute, University of London Press, 1988.

Draelants, Isabelle. "Expérience et autorités dans la philosophie naturelle d'Albert le Grand." In *Expertus sum: L'expérience par les sens dans la philosophie naturelle médiévale*, edited by Thomas Bénatouïl and Isabelle Draelants. Micrologus 40. SISMEL Edizioni del Galluzzo, 2011.

Draper, John William. *History of the Conflict Between Religion and Science*. D. Appleton and Co., 1874.

Druart, Thérèse-Anne. "Al-Kindi's Ethics." *Review of Metaphysics* 47, no. 2 (1993): 329–57.

Du Cange, Charles du Fresne, ed. *Glossarium mediae et infimae latinitatis*. 10 vols. Paris, 1937–1938.

Duhem, Pierre. *Le système du monde: Histoire des doctrines cosmologiques de Platon à Copernic*. 10 vols. Hermann, 1913–59.

———. *To Save the Phenomena: An Essay on the Idea of Physical Theory from Plato to Galileo*. Translated by Edmund Dolan and Chaninah Maschler. University of Chicago Press, 1969.

———. *Sozein ta phainomena, essai sur la notion de théorie physique de Platon à Galilée*. Hermann, 1908.

Dühring, Ernst. "The Two Bacons." Translated by Lydia G. Robinson. *Open Court* 28, no. 8 (August 1914): 468–85.

Dunlop, D. M. "The Translations of al-Bitrīq and Yahyā (Yuhannā) b. al-Batrīq." *Journal of the Royal Asiatic Society of Great Britain and Ireland* 91, no. 3–4 (1959): 140–50.

Eamon, William. *Science and the Secrets of Nature: Books of Secrets in Medieval and Early Modern Culture*. Princeton University Press, 1994.

Eastlake, Charles L. *A History of the Gothic Revival: An Attempt to Show How the Taste for Mediaeval Architecture, Which Lingered in England During the Two Last Centuries, Has Since Been Encouraged and Developed*. Longmans, Green, and Co., 1872.

Easton, Stewart C. *Roger Bacon and His Search for a Universal Science: A Reconsideration of the Life and Work of Roger Bacon in the Light of His Own Stated Purposes*. Columbia University Press, 1952.

Ebke, Almuth and Christoph Haack. "Periodization and Modernity: An Introduction." In "Periodization and Modernity," edited by Almuth Ebke and Christoph Haack. Special issue, *History of European Ideas* 51, no. 2 (2025): 307–20.

Eco, Umberto. *The Name of the Rose*. Translated by William Weaver. Harcourt, 1983.

Edgerton, Samuel Y., Jr. *The Heritage of Giotto's Geometry: Art and Science on the Eve of the Scientific Revolution*. Cornell University Press, 1991.

———. *The Mirror, the Window, and the Telescope: How Renaissance Linear Perspective Changed Our Vision of the Universe*. Cornell University Press, 2009.

Egel, Nikolaus. "Roger Bacon's New Founding of the Sciences in the Thirteenth Century." In *The Philosophy and Science of Roger Bacon: Studies in Honour of Jeremiah Hackett*, edited by Nicola Polloni and Yael Kedar. Routledge, 2021.

Egidi, Pietro, ed. *Codice diplomatico dei Saraceni di Lucera*. L. Pierro et Figlio, 1917.
Ehrenberg, Victor. *Alexander and the Greeks*. Blackwell, 1938.
El Bizri, Nader. "On Vision: Roger Bacon's *Opus maius* (V. 1–2) from the Perspective of Alhazen's *Optics* (1–2)." In *The Sciences of Roger Bacon's* Opus maius, edited by Meagan S. Allen. Routledge, 2025.
Endress, Gerhard. "The Circle of Al-Kindī: Early Arabic Translations from the Greek and the Rise of Islamic Philosophy." In *The Ancient Tradition in Christian and Islamic Hellenism: Studies on the Transmission of Greek Philosophy and Sciences Dedicated to H. J. Drossaart Lulofs on His Ninetieth Birthday*, edited by Gerhard Endress and Remke Kruk. Research School CNWS, 1997.
———. "Die arabischen Übersetzungen von Aristoteles' Schrift 'De caelo.'" PhD diss. Universität Frankfurt-am-Main, 1966.
Erskine, John. *A Pageant of the Thirteenth Century for the Seven Hundredth Anniversary of Roger Bacon*. Plan by John J. Coss. Columbia University Press, 1914.
Esposito, Fernando. "Historicizing Historicism: Reinhard Koselleck and the Periodization of Modernity." In "Periodization and Modernity," edited by Almuth Ebke and Christoph Haack. Special issue, *History of European Ideas* 51, no. 2 (2025): 321–36.
Etherington, Norman. "Barbarians Ancient and Modern." *American Historical Review* 116, no. 1 (2011): 31–57.
Fabian, Johannes. *Time and the Other: How Anthropology Makes Its Object*. Columbia University Press, 2014 [1983].
Farmer, Sharon. "Aristocratic Power and the 'Natural' Landscape: The Garden Park at Hesdin, 1291–1302." *Speculum* 88, no. 3 (2013): 644–80.
Fébvre, Lucien. "How Jules Michelet Invented the Renaissance." In *A New Kind of History: From the Writings of Febvre*, edited by Peter Burke, translated by K. Folca. Routledge and Kegan Paul, 1973.
Feingold, Mordechai. "'Experimental Philosophy': Invention and Rebirth of a Seventeenth-Century Concept." *Early Science and Medicine* 21, no. 1 (2016): 1–28.
Findlen, Paula. "Inventing the Middle Ages: An Early Modern Forger Hiding in Plain Sight." In *For the Sake of Learning: Essays in Honor of Anthony Grafton*, edited by Ann Blair and Anja-Silvia Goering. 2 vols. Brill, 2016.
Firnhaber-Baker, Justine. *House of Lilies: The Dynasty That Made Medieval France*. Basic Books, 2024.
Fránquiz, José Antonio. "The Place of Seneca in the Curriculum of the Middle Ages." In *Arts libéraux et philosophie au Moyen Âge*. Librairie J. Vrin, 1969.
Frederick II. *De arte venandi cum avibus*. Edited by Carl Arnold Willemsen. In aedibus insulae, 1942.
Ganim, John M. *Medievalism and Orientalism*. Palgrave Macmillan, 2005.
Gasper, Giles E. M., Brian K. Tanner, Sigbjørn Sønnesyn, and Nader el-Bizri. "Travelling Optics: Robert Grosseteste and the Optics Behind the Rainbow." In *Medieval Science in the North: Travelling Wisdom, 1000–1500*, edited by Christian Etheridge and Michele Campopiano. Brepols, 2021.
Gasquet, F. A. "An Unpublished Fragment of a Work by Roger Bacon." *English Historical Review* 12, no. 47 (1897): 494–517.
Gaullier-Bougassas, Catherine, ed. *La fascination pour Alexandre le Grand dans les littératures européennes (Xe-XVIe siècle)*. 4 vols. Brepols, 2014.

———. "Savoir scientifique et 'roman historique': L'Alexandre anglo-normand de Thomas de Kent." In *Savoirs et fiction au Moyen Âge et à la Renaissance*, edited by Dominique Boutet and Joëlle Ducos. Presses de l'Université Paris-Sorbonne, 2015.

Geary, Patrick J. *The Myth of Nations: The Medieval Origins of Europe*. Princeton University Press, 2002.

Getz, Faye. "Roger Bacon and Medicine: The Paradox of the Forbidden Fruit and the Secrets of Long Life." In *Roger Bacon and the Sciences: Commemorative Essays*, edited by Jeremiah Hackett. Brill, 1997.

———. "To Prolong Life and Promote Health: Baconian Alchemy and Pharmacy in the English Learned Tradition." In *Health, Disease, and Healing in Medieval Culture*, edited by Sheila D. Campbell, Bert S. Hall, and David N. Klausner. St. Martin's Press, 1992.

Giancoli, Douglas C. *Physics: Principles with Applications*. 7th ed. Pearson, 2014.

Giese, Martina. "The *De Arte venandi cum avibus* of Emperor Frederick II." In *Raptor and Human: Falconry and Bird Symbolism Through the Millennia on a Global Scale*, edited by Karl-Heinz Gersmann and Oliver Grimm. 4 vols. Wachholtz-Murmann, 2018.

———. "Die Tierhaltung am Hof Kaiser Friedrichs II, zwischen Tradition und Innovation." In *Herrschaftsraüme, Herrschaftspraxis und Kommunikation zur Zeit Kaiser Friedrichs II*, edited by Knut Görich, Jan Keupp, and Theo Broekmann. 2 vols. Herbert Utz Verlag, 2008.

Gillin, Edward. "Whewell and Architecture." In *William Whewell: Victorian Polymath*, edited by Lukas M. Verburgt. University of Pittsburgh Press, 2024.

Gingerich, Owen. *The Eye of Heaven: Ptolemy, Copernicus, Kepler*. American Institute of Physics, 1993.

Godefroy, Frédéric, ed. *Dictionnaire de l'ancienne langue française et tous ses dialects du IXe au XVe siècle*. 8 vols. Slatkine Reprints, 1982.

Gómez, Pablo F. *The Experiential Caribbean: Creating Knowledge and Healing in the Early Modern Atlantic*. University of North Carolina Press, 2017.

Gössmann, Elisabeth. *Antiqui und Moderni im Mittelalter: Eine geschichtliche Standortbestimmung*. Ferdinand Schöningh, 1974.

Graeber, David, and David Wengrow. *The Dawn of Everything: A New History of Humanity*. Farrar, Straus and Giroux, 2021.

Grafton, Anthony. *Magus: The Art of Magic from Faustus to Agrippa*. Belknap Press of Harvard University Press, 2023.

Grafton, Anthony, and Megan Williams. *Christianity and the Transformation of the Book: Origen, Eusebius, and the Library of Caesarea*. Belknap Press of Harvard University Press, 2006.

Gransden, Antonia. *Historical Writing in England, c. 550–c. 1307*. 2 vols. Routledge, 1996 [Cornell University Press, 1974].

Grant, Edward. *The Foundations of Modern Science in the Middle Ages*. Cambridge University Press, 1996.

———. *Physical Science in the Middle Ages*. Wiley, 1971.

Green, Monica. *The Trotula: A Medieval Compendium of Women's Medicine*. University of Pennsylvania Press, 2003.

Greenblatt, Stephen. *The Swerve: How the World Became Modern*. W. W. Norton, 2011.

Greene, Robert. *The Most Famous History of the Learned Fryer Bacon: Shewing His Parentage and Birth, How He Came to Be a Scholar and to Study Art-Magick; with the Many*

Wonderful Things He Did in His Life-Time, to the Amazement of the Whole World; in Making a Brazen Head, to Have Walled All England with Brass: With His Penitent Death, Also, The Merry Waggeries of His Man Miles: And the Exploits of Vandermaster, a German, and Fryer Bungy, an English Conjurer; With the Manner of Their Woful Deaths, as a Warning to Others: Being All Very Profitable and Pleasant to the Reader. Edited by Fryer Bungy. Thomas Norris, Sign of the Looking-Glass on London Bridge, [1720?].

Grignaschi, Mario. "La figure d'Alexandre chez les Arabes et sa genèse." *Arabic Sciences and Philosophy* 3, no. 2 (1993): 205–34.

———. "Les *Rasā'il 'Aristātālīsa 'ilā-l-Iskandar* de Sālim Abū-l-'Alā' et l'activité culturelle a l'époque omayyade." *Bulletin d'études orientales de Damas* 19 (1965): 7–83.

———. "L'origine et les métamorphoses du *Sirr al-asrār*." *Archives d'histoire doctrinale et littéraire du moyen âge* 43 (1976): 7–112.

———. "Remarques sur la formation et l'interprétation du *Sirr al-asrār*." In *Pseudo-Aristotle, The Secret of Secrets: Sources and Influences*, edited by W. F. Ryan and Charles B. Schmitt. Warburg Institute, University of London, 1982.

Grosseteste, Robert. *Commentarius in Posteriorum analyticorum libros.* Edited by Pietro Rossi. Unione Accademica Nazionale. Leo S. Olschki, 1981.

———. *De cometis.* In *Die philosophischen Werke des Robert Grosseteste, Bischofs von Lincoln*, edited by Ludwig Baur. Aschendorff Verlag, 1912.

Gunderson, Lloyd L., ed. and trans. *Alexander's Letter to Aristotle About India.* Hain, 1980.

Gutas, Dimitri. "Ethische Schriften im Islam." In *Orientalisches Mittelalter*, edited by Wolfhart Heinrichs. AULA-Verlag, 1990.

———. *Greek Thought, Arabic Culture: The Graeco-Arabic Translation Movement in Baghdad and Early 'Abbāsid Society (2nd–4th/8th–10th Centuries).* Routledge, 1998.

Hackett, Jeremiah. "Aristotle, *Astrologia*, and Controversy at the University of Paris (1266–1274)." In *Learning Institutionalized: Teaching in the Medieval University*, edited by John Van Engen. University of Notre Dame Press, 2000.

———. "The Attitude of Roger Bacon to the *Scientia* of Albertus Magnus." In *Albertus Magnus and the Sciences, Commemorative Essays 1980*, edited by James A. Weisheipl. Pontifical Institute of Mediaeval Studies, 1980.

———. "*Ego Expertus Sum*: Roger Bacon's Science and the Origins of Empiricism." In *Expertus sum: L'expérience par les sens dans la philosophie naturelle médiévale*, edited by Thomas Bénatouïl and Isabelle Draelants. Micrologus 40. SISMEL Edizioni del Galluzzo, 2011.

———. "From *Sapientes antiqui* at Lincoln to the New *Sapientes moderni* at Paris c. 1260–1280: Roger Bacon's Two Circles of Scholars." In *Robert Grosseteste and the Pursuit of Religious and Scientific Learning in the Middle Ages*, edited by Jack P. Cunningham and Mark Hocknull. Studies in the History of Philosophy of Mind 18. Springer, 2016.

———. "Moral Philosophy and Rhetoric in Roger Bacon." *Philosophy and Rhetoric* 20, no. 1 (1987): 18–40.

———, ed. *Roger Bacon and the Sciences: Commemorative Essays.* Brill, 1997.

———. "Roger Bacon, Aristotle, and the Parisian Condemnations of 1270, 1277." *Vivarium* 35, no. 2 (1997): 283–314.

———. "Roger Bacon: His Life, Career, and Works." In *Roger Bacon and the Sciences: Commemorative Essays*, edited by Jeremiah Hackett. Brill, 1997.

———. "Roger Bacon on *Scientia Experimentalis*." In *Roger Bacon and the Sciences: Commemorative Essays*, edited by Jeremiah Hackett. Brill, 1997.

———. "Roger Bacon's Concept of Experience: A New Beginning in Medieval Philosophy?" *Modern Schoolman* 86, no. 1/2 (November 2008): 123–46.

———. "Roger Bacon's Various Texts on *Perspectiva* in His Later Works: The Significance of Two New Versions of His *Perspectiva*." In *Arabic and Latin Theory of Perspective*, edited by Agostino Paravicini Bagliani. Micrologus 29. SISMEL Edizioni del Galluzzo, 2021.

———. "Roger Bacon's *Scientia Experimentalis*: A Review of Bacon's Diverse Treatises." In *The Sciences of Roger Bacon's* Opus maius, edited by Meagan S. Allen. Routledge, 2025.

———. "*Scientia Experimentalis*: From Robert Grosseteste to Roger Bacon." In *Robert Grosseteste: New Perspectives on His Thought and Scholarship*, edited by James McEvoy. Brepols, 1995.

Halleux, Robert. "Entre philosophie naturelle et savoir d'ingénieur: *L'Epistola de magnete* de Pierre de Maricourt." *Archives Internationales d'Histoire des Sciences* 56, no. 156–57 (2006): 3–17.

Hanegraaff, Wouter J. *Esotericism and the Academy: Rejected Knowledge in Western Culture.* Cambridge University Press, 2012.

Hankins, James. *Virtue Politics: Soulcraft and Statecraft in Renaissance Italy.* Belknap Press of Harvard University Press, 2019.

Hartog, François. *Régimes d'historicité: Présentisme et expériences du temps.* Seuil, 2003.

———. *Regimes of Historicity: Presentism and Experiences of Time.* Translated by Saskia Brown. Columbia University Press, 2015.

Haskins, Charles Homer. *Renaissance and Renewal in the Twelfth Century.* Edited by Robert L. Benson, Giles Constable, and Carol D. Lanham. Harvard University Press, 1982.

———. *The Renaissance of the Twelfth Century.* Harvard University Press, 1927.

Hasse, Dag Nikolaus. *Success and Suppression: Arabic Sciences and Philosophy in the Renaissance.* Harvard University Press, 2016.

Hava, J. G., ed. *Al-Faraid Arabic-English Dictionary.* Catholic Press, 1964.

Hecht, Eugene. *Optics.* 5th ed. Pearson, 2016.

Hegel, Georg Wilhelm Friedrich. *Lectures on the History of Philosophy.* Translated by E. S. Haldane and F. H. Simson. 3 vols. London, 1892–96.

Herde, Peter. "Literary Activities of the Imperial and Papal Chanceries During the Struggle Between Frederick II and the Papacy." In *Intellectual Life at the Court of Frederick II Hohenstaufen*, edited by William Tronzo. National Gallery of Art, 1994.

Hochberg, Herbert. "The Empirical Philosophy of Roger and Francis Bacon." *Philosophy of Science* 20, no. 4 (1953): 313–26.

Hodgson, Marshall G. S. *The Venture of Islam: Conscience and History in a World Civilization.* 3 vols. University of Chicago Press, 1974.

Holsinger, Bruce. *Neomedievalism, Neoconservatism, and the War on Terror.* University of Chicago Press, 2007.

Hovdaugen, Even. "'Una et eadem': Some Observations on Roger Bacon's Greek Grammar." In *De ortu grammaticae: Studies in Medieval Grammar and Linguistic Theory in Memory of Jan Pinborg*, edited by Geoffrey L. Bursill-Hall, Sten Ebbesen, and E. F. Konrad Koerner. John Benjamins Pub. Co., 1990.

Huff, Toby E. *Intellectual Curiosity and the Scientific Revolution: A Global Perspective.* Cambridge University Press, 2010.

———. *The Rise of Early Modern Science: Islam, China, and the West.* 2nd ed. Cambridge University Press, 2003.

Hughes, Jonathan. "The Humanity of Thomas Charnock, an Elizabethan Alchemist." In *Mystical Metal of Gold: Essays on Alchemy and Renaissance Culture*, edited by Stanton J. Linden. AMS Press, 2007.

Ibn Manzur, ed. *Lisān Al-Arab*. 18 vols. Dar Sader Publishers, 2017.

Ibn Munqidh, Usāma. *Kitāb al- Iʿtibār* (*The Book of Contemplation*), edited and translated by Paul M. Cobb. Smithsonian Libraries, 2008.

Ilardi, Vincent. *Renaissance Vision from Spectacles to Telescopes*. American Philosophical Society, 2007.

Ingham, Patricia Clare. *The Medieval New: Ambivalence in an Age of Innovation*. University of Pennsylvania Press, 2015.

Ingham, Patricia Clare, and Michelle R. Warren, eds. *Postcolonial Moves: Medieval Through Modern*. Palgrave Macmillan, 2003.

Inglebert, Hervé. "The Universal Chronicle in Antiquity and in the Middle Ages." In *Vehicles of Transmission, Translation, and Transformation in Medieval Textual Culture*, edited by Robert Wisnovsky, Faith Wallis, Jamie C. Fumo, and Carlos Fraenkel. Brepols, 2011.

Isidore of Seville. *Etymologiarum*. Edited by W. M. Lindsay. 2 vols. Clarendon Press, 1911.

Jacquart, Danielle. "Le physiognomie à l'époque de Frédéric II: Le traité de Michel Scot." In *Recherches médiévales sur la nature humaine: Essais sur la réflexion médicale (XIIe–XVe s.)*. Micrologus 63. SISMEL Edizioni del Galluzzo, 1994.

Jaritz, Gerhard, and Gerson Moreno-Riaño, eds. *Time and Eternity: The Medieval Discourse*. Brepols, 2003.

Johnson, Timothy J. "'Wisdom Has Built Her House; She Has Set Up Her Seven Pillars': Roger Bacon, Franciscan Wisdom, and Conversion to the Sciences." In *The English Province of the Franciscans (1224–c.1350)*, edited by Michael J. P. Robson. Brill, 2017.

Jordan, William Chester. *The French Monarchy and the Jews: From Philip Augustus to the Last Capetians*. University of Pennsylvania Press, 1989.

Jourdain, Charles. "Discussion de quelques points de la biographie de Roger Bacon." In *Excursions historiques et philosophiques à travers le moyen âge*. Firmin-Didot, 1888.

Kamp, Norbert. *Enciclopedia dei Papi*. 3 vols. Istituto della Enciclopedia Italiana, 2000.

Kassell, Lauren. "'All Was This Land Full Fill'd of Faerie,' or Magic and the Past in Early Modern England." *Journal of the History of Ideas* 67, no. 1 (2006): 107–22.

Kavenna, Joanna. *A Field Guide to Reality*. Riverrun, 2017.

Kelly, Douglas. *The Art of Medieval French Romance*. University of Wisconsin Press, 1992.

Kemp, Martin. *The Science of Art: Optical Themes in Western Art from Brunelleschi to Seurat*. Yale University Press, 1990.

———. *Visions of Heaven: Dante and the Art of Divine Light*. Lund Humphries, 2021.

Kennedy, Philip F., ed. *On Fiction and Adab in Medieval Arabic Literature*. Harrassowitz Verlag, 2005.

Khalaf, Omar, ed. *Alexander and Dindimus*. Universitätsverlag Winter, 2017.

Kheirandish, Elaheh. "The Footprints of 'Experiment' in Early Arabic Optics." *Early Science and Medicine* 14 (2009): 79–104.

———. "The Mixed Mathematical Sciences: Optics and Mechanics in the Islamic Middle Ages." In *Cambridge History of Science*, vol. 2, *Medieval Science*, edited by David C. Lindberg and Michael H. Shank. Cambridge University Press, 2013.

Kieckhefer, Richard. *Magic in the Middle Ages*. 3rd ed. Cambridge University Press, 2021.

Kleinert, Andreas. "Le moteur magnéto-mécanique de Pierre de Maricourt: Comptes-rendus du séminaire; Origine des idées scientifiques, ruptures et continuités." *Centre Commun d'Histoire des Sciences et d'Epistémologie de Lille* 1 (2004–5): 22–34.

Koselleck, Reinhart. *Futures Past: On the Semantics of Historical Time*. Translated by Keith Tribe. Columbia University Press, 2004.

Koyré, Alexandre. *Études Galiléennes*. 3 vols. Hermann, 1939.

———. *From the Closed World to the Infinite Universe*. Johns Hopkins University Press, 1957.

———. "Galileo and Plato." *Journal of the History of Ideas* 4, no. 4 (1943): 400–428.

———. "The Origins of Modern Science: A New Interpretation." *Diogenes* 4, no. 16 (1956): 1–22.

Kuhn, Thomas S. *The Structure of Scientific Revolutions*. 4th ed. University of Chicago Press, 2012 [1962].

Latour, Bruno. *Nous n'avons jamais été modernes: Essai d'anthropologie symétrique*. La Découverte, 1991.

———. *We Have Never Been Modern*. Translated by Catherine Porter. Harvard University Press, 1993.

Leff, Gordon. *Paris and Oxford Universities in the Thirteenth and Fourteenth Centuries: An Institutional and Intellectual History*. John Wiley and Sons, 1968.

Lemay, Richard. "Roger Bacon's Attitude Toward the Latin Translations and Translators of the Twelfth and Thirteenth Centuries." In *Roger Bacon and the Sciences: Commemorative Essays*, edited by Jeremiah Hackett. Brill, 1997.

Leo. "Die interpolierte Historia de preliis." In *Der Alexanderroman des Archipresbyters Leo*, edited by Friedrich Pfister. C. Winter, 1913.

Leong, Elaine. *Recipes and Everyday Knowledge: Medicine, Science, and the Household in Early Modern England*. University of Chicago Press, 2018.

Levine, Philippa. *The Amateur and the Professional: Antiquarians, Historians and Archaeologists in Victorian England, 1838–1886*. Cambridge University Press, 1986.

Libera, Alain de. "Roger Bacon et la logique." In *Roger Bacon and the Sciences: Commemorative Essays*, edited by Jeremiah Hackett. Brill, 1997.

Lindberg, David C. *The Beginnings of Western Science: The European Scientific Tradition in Philosophical, Religious, and Institutional Context, Prehistory to A.D. 1450*. 2nd ed. University of Chicago Press, 1992.

———. "Lines of Influence in Thirteenth-Century Optics: Bacon, Witelo, and Pecham." *Speculum* 46, no. 1 (1971): 66–83.

———, ed. *Roger Bacon and the Origins of Perspectiva in the Middle Ages: A Critical Edition and English Translation of Bacon's Perspectiva with Introduction and Notes*. Clarendon Press, 1996.

———. "Roger Bacon on Light, Vision and the Universal Emanation of Force." In *Roger Bacon and the Sciences: Commemorative Essays*, edited by Jeremiah Hackett. Brill, 1997.

———, ed. *Roger Bacon's Philosophy of Nature: A Critical Edition, with English Translation, Introduction, and Notes, of De Multiplicatione Specierum and De Speculis Comburentibus*. Clarendon Press, 1983.

———. "Roger Bacon's Theory of the Rainbow: Progress or Regress?" *Isis* 57, no. 2 (1966): 235–48.

———. "Science as Handmaiden: Roger Bacon and the Patristic Tradition." *Isis* 78, no. 4 (1987): 518–36.

———. *Theories of Vision from Al-Kindi to Kepler*. University of Chicago Press, 1976.

———. "The Western Reception of Arabic Optics." In *The Encyclopedia of the History of Arabic Science*, vol. 3, *Technology, Alchemy and Life Sciences*, edited by Roshdi Rashed and Régis Morelon. 3 vols. Routledge, 1996.

Lindberg, David C., and Ronald L. Numbers. "Beyond War and Peace: A Reappraisal of the Encounter Between Christianity and Science." *Perspectives on Science and Christian Faith* 39, no. 3 (1987): 140–49.

Lipton, Sara. *Images of Intolerance: The Representation of Jews and Judaism in the Bible Moralisée.* University of California Press, 1999.

Little, A. G., ed. "On Roger Bacon's Life and Works." In *Roger Bacon: Essays Contributed by Various Writers on the Occasion of the Commemoration of the Seventh Centenary of His Birth*, edited by A. G. Little. Clarendon Press, 1914.

———. *Roger Bacon: Essays Contributed by Various Writers on the Occasion of the Commemoration of the Seventh Centenary of His Birth.* Clarendon Press, 1914.

———. "Roger Bacon's Works with References to the MSS. and Printed Editions." In *Roger Bacon: Essays Contributed by Various Writers on the Occasion of the Commemoration of the Seventh Centenary of His Birth*, edited by A. G. Little. Clarendon Press, 1914.

Liudprand of Cremona. *The Embassy to Constantinople and Other Writings.* Edited by John Julius Norwich. Translated by F. A. Wright. J. M. Dent, 1993.

Livingston, Julie. *Self-Devouring Growth: A Planetary Parable as Told from Southern Africa.* Duke University Press, 2019.

Long, Pamela O. *Artisan/Practitioners and the Rise of the New Sciences, 1400–1600.* University of Oregon, 2011.

———. *Technology in Mediterranean and European Lands, 600–1600.* Johns Hopkins University Press, 2025.

Lowden, John. *The Making of the Bibles Moralisées.* 2 vols. Penn State University Press, 2000.

Lyon, Gregory B. "Baudouin, Flacius, and the Plan for the Magdeburg Centuries." *Journal of the History of Ideas* 64, no. 2 (2003): 253–72.

Maddicott, J. R. *Simon de Montfort.* Cambridge University Press, 1994.

Maddox, Donald, and Sara Sturm-Maddox, eds. *The Medieval French Alexander.* State University of New York Press, 2002.

Mallette, Karla. *European Modernity and the Arab Mediterranean: Toward a New Philology and a Counter-Orientalism.* University of Pennsylvania Press, 2010.

Maloney, Thomas S. "Introduction." In *Roger Bacon, Compendium of the Study of Theology*, edited and translated by Thomas S. Maloney. Brill, 1988.

Manix, Pam. "Oxford: Mapping the Medieval Jewry." In *The Jews of Europe in the Middle Ages (Tenth to Fifteenth Centuries)*, edited by Christoph Cluse. Brepols, 2004.

Mantovani, Mattia. "The First of All Natural Sciences: Roger Bacon on *Perspectiva* and Human Knowledge." *Vivarium* 59 (2021): 186–214.

Manzalaoui, Mahmoud. "The Pseudo-Aristotelian *Kitāb Sirr al-asrār*: Facts and Problems." *Oriens* 23–24 (1974): 147–257.

Marrone, Steven. "Metaphysics and Science in the Thirteenth Century: William of Auvergne, Robert Grosseteste, and Roger Bacon." In *Routledge History of Philosophy*, vol. 3, *Medieval Philosophy*, edited by John Marenbon. Routledge, 1998.

Martin, Craig. "Rethinking Renaissance Averroism." *Intellectual History Review* 17, no. 1 (2007): 3–19.

———. *Subverting Aristotle: Religion, History, and Philosophy in Early Modern Science.* Johns Hopkins University Press, 2014.

Martin, George R. R. *A Game of Thrones*. Bantam, 1996.

Massa, Eugenio. *Ruggero Bacone etica e poetica nella storia dell' "Opus maius."* Edizioni di Storia e Letteratura, 1955.

Matton, Sylvain. "An Irradiation of Latin Grammarians; or, The *De radiis* Is Not by al-Kindī." *Magic, Ritual, and Witchcraft* 17, no. 3 (2022): 437–55.

Matus, Zachary A. *Franciscans and the Elixir of Life: Religion and Science in the Later Middle Ages*. University of Pennsylvania Press, 2017.

———. "Reconsidering Roger Bacon's Apocalypticism in Light of His Alchemical and Scientific Thought." *Harvard Theological Review* 105, no. 2 (2012): 189–222.

McCallum, R. Ian. "Alchemical Scrolls Associated with George Ripley." In *Mystical Metal of Gold: Essays on Alchemy and Renaissance Culture*, edited by Stanton J. Linden. AMS Press, 2007.

McKitterick, Rosamond. *History and Memory in the Carolingian World*. Cambridge University Press, 2004.

McVaugh, Michael. "Determining a Drug's Properties: Medieval Experimental Protocols." *Bulletin of the History of Medicine* 91, no. 2 (2017): 183–209.

Melvin-Koushki, Matthew. "Taḥqīq vs. Taqlīd in the Renaissances of Western Early Modernity." *Philological Encounters* 3, no. 1–2 (2018): 193–249.

Menand, Louis. *The Marketplace of Ideas: Reform and Resistance in the American University*. W. W. Norton, 2010.

Menocal, María Rosa. *The Arabic Role in Medieval Literary History: A Forgotten Heritage*. University of Pennsylvania Press, 1990.

Michelet, Jules. *Histoire de France*. 19 vols. Hachette, 1831–67.

The Mirror of Alchimy: Composed by the Thrice-Famous and Learned Fryer, Roger Bachon. Edited by Stanton J. Linden. Garland Publishing, 1992.

Miyashiro, Adam. "Our Deeper Past: Race, Settler Colonialism, and Medieval Heritage Politics." *Literature Compass* 16, no. 9–10 (2019). https://doi.org/10.1111/lic3.12550.

Molland, A. George. "Roger Bacon as Magician." *Traditio* 30 (1974): 445–60.

———. "Roger Bacon's Appropriation of Past Mathematics." In *Tradition, Transmission, Transformation: Proceedings of Two Conferences on Pre-modern Science Held at the University of Oklahoma*, edited by F. Jamil Ragep, Sally P. Ragep, and Steven John Livesey. Brill, 1996.

———. "Roger Bacon's Knowledge of Mathematics." In *Roger Bacon and the Sciences: Commemorative Essays*, edited by Jeremiah Hackett. Brill, 1997.

Mommsen, Theodore E. "Petrarch's Conception of the 'Dark Ages.'" *Speculum* 17, no. 2 (1942): 226–42.

Mora-Lebrun, Francine. *"Metre en romanz": Les romans d'Antiquité du XIIème siècle et leur postérité (XIIIème-XIVème siècle)*. Honoré Champion, 2008.

Morpurgo, Piero. "*Philosophia Naturalis* at the Court of Frederick II: From the Theological Method to the *Ratio Secundum Physicam* in Michael Scot's *De anima*." In *Intellectual Life at the Court of Frederick II Hohenstaufen*, edited by William Tronzo. Studies in the History of Art 44. National Gallery of Art, 1994.

Morton, Jonathan. "*Engin*: Creativity, Invention, and *Knowledge* [*sic*] in the Medieval Romance Tradition of Alexander the Great." *Romanic Review* 111, no. 2 (2020): 205–24.

Mukharji, Projit Bihari. *Doctoring Traditions: Ayurveda, Small Technologies, and Braided Sciences*. University of Chicago Press, 2016.

Murdoch, John E. "Philosophy and the Enterprise of Science in the Later Middle Ages." In *The Interaction Between Science and Philosophy*, edited by Yehuda Elkana and Samuel Sambursky. Humanities Press, 1974.

———. "Pierre Duhem and the History of Late Medieval Science and Philosophy in the Latin West." In *Gli studi di filosofia medievale tra Otto e Novecento*, edited by Imbach Ruedi and Maierù Alfonso. Edizioni di Storia e Letteratura, 1991.

Murray, Alexander. "Should the Middle Ages Be Abolished?" *Essays in Medieval Studies* 21 (2004): 1–22.

Myers, Constance. "Paul Carus and *The Open Court*: History of a Journal." *Midcontinent American Studies Journal* 5, no. 2 (1964): 57–68.

Nagel, Alexander, and Christopher S. Wood. *Anachronic Renaissance*. Zone Books, 2010.

Naudé, Gabriel. *Apologie pour tous les grandes personnages qui ont esté faussement soupçonnez de magie*. François Targa, 1625.

———. *The History of Magick by Way of Apology, for All the Wise Men Who Have Unjustly Been Reputed Magicians, from the Creation, to the Present Age*. Translated by John Davies. London, 1657.

Newman, William R. "The Alchemy of Roger Bacon and the *Tres Epistolae* Attributed to Him." In *Comprendre et maîtriser la nature Au Moyen Age: Mélanges d'histoire des sciences offerts à Guy Beaujouan*, edited by Danielle Jacquart. Droz, 1994.

———. "An Overview of Roger Bacon's Alchemy." In *Roger Bacon and the Sciences: Commemorative Essays*, edited by Jeremiah Hackett. Brill, 1997.

———. "The Philosopher's Egg: Theory and Practice in the Alchemy of Roger Bacon." *Micrologus* 3 (1995): 75–101.

———. "Roger Bacon and Secrecy: Alchemical Techniques of Deception and the *Doctor Mirabilis*." In *The Sciences of Roger Bacon's* Opus maius, edited by Meagan S. Allen. Routledge, 2025.

Nirenberg, David. *Anti-Judaism: The Western Tradition*. Norton, 2013.

Nolan, Edmond, and Samuel Abraham Hirsch, eds. *The Greek Grammar of Roger Bacon and a Fragment of His Hebrew Grammar*. Cambridge University Press, 1902.

North, John D. *God's Clockmaker: Richard of Wallingford and the Invention of Time*. Hambledon, 2005.

———. "Monasticism and the First Mechanical Clocks." In *The Study of Time II: Proceedings of the Second Conference of the International Society for the Study of Time, Lake Yamanaka, Japan*, edited by J. T. Fraser and Nathaniel Lawrence. Springer Verlag, 1975.

Nothaft, C. Philipp E. "Duking It Out in the Arena of Time: Chronology and the Christian-Jewish Encounter (1100–1600)." *Medieval Encounters* 22, no. 1–3 (2016): 213–35.

Nummedal, Tara. *Anna Zieglerin and the Lion's Blood: Alchemy and End Times in Reformation Germany*. University of Pennsylvania Press, 2019.

O'Keeffe, D. "Roger Bacon." *Irish Ecclesiastical Record*, September 1914, 346–57.

Oliver, Simon. "Robert Grosseteste on Light, Truth and *Experimentum*." *Vivarium* 42, no. 2 (2004): 151–80.

Omar, Saleh Beshara. *Ibn al-Haytham's Optics: A Study of the Origins of Experimental Science*. Bibliotheca Islamica, 1977.

The Open Court: Devoted to the Science of Religion, the Religion of Science, and the Extension of the Religious Parliament Idea, August 1918.

Orme, Nicholas. *Education in the West of England, 1066–1548: Cornwall, Devon, Dorset, Gloucestershire, Somerset, Wiltshire.* University of Exeter Press, 1976.

———. *Medieval Schools: From Roman Britain to Renaissance England.* Yale University Press, 2006.

Owens, James A. *Here, There Be Dragons.* Simon and Schuster, 2010.

Oxford English Dictionary. 2nd ed. Oxford University Press. https://www.oed.com.

Panti, Cecilia. "In the Margins of the *Posterior Analytics*, Robert Grosseteste and the 'Latin Philoponus.'" *Revista Española Filosofía Medieval* 30, no. 1 (2023): 15–52.

———. *Moti, virtù e motori celesti nella cosmologia di Roberto Grossatesta.* SISMEL Edizioni del Galluzzo, 2001.

———. "The Theological Use of Science in Robert Grosseteste and Adam Marsh According to Roger Bacon: The Case Study of the Rainbow." In *Robert Grosseteste and the Pursuit of Religious and Scientific Learning in the Middle Ages*, edited by Jack P. Cunningham and Mark Hocknull. Springer, 2016.

Paravicini Bagliani, Agostino, ed. *Medicina e scienze della natura, alla corte dei papi nel Duecento.* Centro italiano di studi sull'alto medioevo, 1991.

———. *The Pope's Body.* Translated by David S. Peterson. University of Chicago Press, 2000.

———. "Ruggero Bacone autore del *De retardatione accidentium senectutis*?" *Studi medievali* 28, no. 2 (1987): 707–27.

———. "Ruggero Bacone, Bonifacio VIII, e la teoria della *prolongatio vitae*." In *Medicina e scienza della natura, alla corte dei papi nel Duecento*, edited by Agostino Paravicini Bagliani. Centro italiano di studi sull'alto medioevo, 1991.

———. "Ruggero Bacone e l'alchimia di lunga vita: Riflessioni su il testi." In *Alchimia e medicina nel medioevo*, edited by Chiara Crisciani and Agostino Paravicini Bagliani. SISMEL Edizioni del Galluzzo, 2003.

———. "Witelo et la science de l'optique à la cour pontificale de Viterbe (1277)." *Mélanges de l'école française de Rome (moyen-âge-temps modernes)* 87, no. 2 (1975): 425–53.

Paravicini Bagliani, Agostino, and Steven J. Williams. "Addendum." *Studi medievali* 28, no. 2 (1987): 727–28.

Park, Katharine. "Myth: That the Medieval Church Prohibited Human Dissection." In *Galileo Goes to Jail, and Other Myths About Science and Religion*, edited by Ronald L. Numbers. Harvard University Press, 2009.

———. *Secrets of Women: Gender, Generation, and the Origins of Human Dissection.* Zone, 2006.

Park, Katharine, and Lorraine Daston. "The Age of the New." In *Cambridge History of Science*, vol. 3, *Early Modern Science.* Cambridge University Press, 2006.

Parry, Glyn. *The Arch Conjuror of England: John Dee.* Yale University Press, 2011.

Patterson, Lee. "Critical Historicism and Medieval Studies." In *Literary Practice and Social Change in Britain, 1380–1530*, edited by Lee Patterson. University of California Press, 1990.

———, ed. *Literary Practice and Social Change in Britain, 1380–1530.* University of California Press, 1990.

Payne, Alina. "Vasari, Architecture, and the Origins of Historicizing Art." *RES: Anthropology and Aesthetics* 40 (2001): 51–76.

Pearsall, Derek. "Hoccleve's *Regiment of Princes*: The Poetics of Royal Self-Representation." *Speculum* 69 (1994): 386–410.

Pedersen, Olaf. *The First Universities: Studium Generale and the Origins of University Education in Europe.* Translated by Richard North. Cambridge University Press, 1997.

Peter Peregrinus of Maricourt. "Epistola de magnete." In *Petrus Peregrinus Opera*, edited by Loris Sturlese and Ron B. Thomson. Centro di cultura medievale della Scuola Normale Superiore, 1995.

———. *The Letter of Petrus Peregrinus on the Magnet, A.D. 1269.* Translated by Fr. Arnold. McGraw, 1904.

Petraitis, Casimir. *The Arabic Version of Aristotle's "Meteorology".* Dar el-Machreq / Librarie orientale, 1967.

Plot, Robert. *The Natural History of Oxford-shire, Being an Essay Towards the Natural History of England.* Oxford, 1677.

Pocock, J. G. A. *Barbarism and Religion.* Vol. 3, *The First Decline and Fall.* Cambridge University Press, 2003.

Polloni, Nicola, and Yael Kedar. "Disentangling Roger Bacon's Criticism of Medieval Translations." In *Early Thirteenth-Century English Franciscan Thought*, edited by Lydia Schumacher. De Gruyter, 2021.

———, eds. *The Philosophy and Science of Roger Bacon: Studies in Honour of Jeremiah Hackett.* Routledge, 2021.

Poskett, James. *Horizons: The Global Origins of Modern Science.* Mariner Books, 2022.

Power, Amanda. "Franciscan Advice to the Papacy in the Middle Ages." *History Compass* 5, no. 5 (2007): 1550–75.

———. "Going Among the Infidels: The Mendicant Orders and Louis IX's First Mediterranean Campaign." *Mediterranean Historical Review* 25, no. 2 (2010): 187–202.

———. "A Mirror for Every Age: The Reputation of Roger Bacon." *English Historical Review* 121, no. 492 (2006): 657–92.

———. *Roger Bacon and the Defence of Christendom.* Cambridge University Press, 2011.

———. "Seeking Remedies for Great Danger: Contemporary Appraisals of Roger Bacon's Expertise." In *Knowledge, Discipline, and Power in the Middle Ages: Essays in Honour of David Luscombe*, edited by Joseph Canning, Edmund J. King, and Martial Staub. Brill, 2011.

Quinn, Aleta. "William Whewell's Philosophy of Architecture and the Historicization of Biology." *Studies in History and Philosophy of Science Part C: Studies in History and Philosophy of Biological and Biomedical Sciences* 59 (2016): 11–19.

Rabasa, José. "Decolonizing Medieval Mexico." In *Medievalisms in the Postcolonial World: The Idea of "the Middle Ages" Outside Europe*, edited by Kathleen Davis and Nadia Altschul. Johns Hopkins University Press, 2009.

Ragab, Ahmed. "Translation and the Making of a Medical Archive: The Case of the Islamic Translation Movement." *Osiris* 37, no. 1 (2022): 25–46.

Ragep, F. Jamil. "Islamic Culture and the Natural Sciences." In *Cambridge History of Science*, vol. 2, *Medieval Science*, edited by David C. Lindberg and Michael H. Shank. Cambridge University Press, 2013.

Rambaran-Olm, Mary, and Erik Wade. "What's in a Name? The Past and Present Racism in 'Anglo-Saxon' Studies." In "Literature to 1200," edited by Joshua Davies and Clare A. Lees. Special issue, *Yearbook of English Studies*, 52, no. 1 (2022): 135–53.

Rampling, Jennifer M. *The Experimental Fire: Inventing English Alchemy, 1300–1700.* University of Chicago Press, 2020.

Rankin, Alisha. *The Poison Trials: Wonder Drugs, Experiment, and the Battle for Authority in Renaissance Science*. University of Chicago Press, 2021.

Rashed, Roshdi, ed. *Les mathématiques infinitésimales du IXe au Xie siècle*. Vol. 2. Al-Furquān, 1993.

Raynaud, Christiane "Aristote dans les enluminures du XIIIe siècle." In *L'unité de la culture européenne au Moyen Âge*, edited by Danielle Buschinger and Wolfgang Spiewok. Reineke-Verlag, 1994.

Rey, Alain, ed. *Dictionnaire historique de la langue française*. 2 vols. Le Robert, 1992.

Rivera, José Luis. "Pierre de Maricourt." In *A Companion to Philosophy in the Middle Ages*, edited by Jorge J. E. Garcia and Timothy B. Noone. Blackwell, 2003.

Robinson, Cynthia. *In Praise of Song: The Making of Courtly Culture in Al-Andalus and Provence, 1005–1134 A.D.* Brill, 2002.

Roest, Bert. *Reading the Book of History: Intellectual Contexts and Educational Functions of Franciscan Historiography 1226–ca.1350*. Rijksuniversiteit, 1996.

"Roger Bacon." *British Medical Journal* (June 20, 1914): 1365–67.

"The Roger Bacon Commemoration at Oxford." *Nature* 93, no. 2327 (June 4, 1914): 354–55.

Rollo-Koster, Joëlle, and Thomas M. Izbicki, eds. *A Companion to the Great Western Schism (1378–1417)*. Brill, 2009.

Rosier-Catach, Irène. "Roger Bacon and Grammar." In *Roger Bacon and the Sciences: Commemorative Essays*, edited by Jeremiah Hackett. Brill, 1997.

Ross, David J. A. *Studies in the Alexander Romance*. Pindar Press, 1985.

Rubin, Patricia Lee. *Giorgio Vasari: Art and History*. Yale University Press, 1995.

Rubruck, William. *The Journey of William of Rubruck to the Eastern Parts of the World, 1253–1255*. Edited and translated by William Woodville Rockhill. Hakluyt Society, 1900.

Rutkin, H. Darrel. "An Idealized Astrological Courtier at a 13th Century Papal and Royal Court: The Case of Roger Bacon." In *De Frédéric II à Rodolphe II. Astrologie, divination et magie dans les cours (XIIIe-XVIIe siècle)*, edited by Jean-Patrice Boudet, Martine Ostorero, and Agostino Paravicini Bagliani. Micrologus 85. SISMEL Edizioni del Galluzzo, 2017.

———. "The 'Unnamed Master' Revisited." In *The Sciences of Roger Bacon's* Opus maius, edited by Meagan S. Allen. Routledge, 2025.

Ryden, Barbara, and Bradley M. Peterson. *Foundations of Astrophysics*. Addison-Wesley, 2010.

Sabra, Abdelhamid I. "The Appropriation and Subsequent Naturalization of Greek Science in Medieval Islam: A Preliminary Statement." *History of Science* 25, no. 3 (1987): 223–43.

———. "The 'Commentary' That Saved the Text: The Hazardous Journey of Ibn al-Haytham's Arabic *Optics*." *Early Science and Medicine* 12, no. 2 (2007): 117–33.

———. "Ibn al-Haytham." In *Dictionary of Scientific Biography*, edited by Charles Coulston Gillispie. Charles Scribner's Sons, 1972

———. "Ibn Al-Haytham's Revolutionary Project in Optics: The Achievement and the Obstacle." In *The Enterprise of Science in Islam: New Perspectives*, edited by Abdelhamid I. Sabra and Jan P. Hogendijk. MIT Press, 2003.

———. "One Ibn al-Haytham or Two? An Exercise in Reading the Bio-Bibliographic Sources." *Zeitschrift für Geschichte der Arabisch-Islamischen Wissenschaften* 12 (1998): 1–50.

———. *The Optics of Ibn al-Haytham: Books I–III, On Direct Vision*. 2 vols. Warburg Institute, University of London, 1989.

Said, Edward W. *Orientalism*. Vintage, 1979.

Saliba, George. *Islamic Science and the Making of the European Renaissance.* MIT Press, 2007.

Sandys, John Edwin. "Roger Bacon." *Homiletic Review: An International Magazine of Religion, Theology, and Philosophy,* June 1914, 433–38.

———. "Roger Bacon." *Saturday Review of Politics, Literature, Science, and Art,* June 1914.

Sankovitch, Anne-Marie. "The Myth of the 'Myth of the Medieval': Gothic Architecture in Vasari's *Rinascita* and Panofsky's *Renaissance.*" *RES: Anthropology and Aesthetics* 40 (2001): 29–50.

Sannino, Antonella. "Guillaume d'Auvergne e i *libri experimentorum.*" In *Expertus sum: L'expérience par les sens dans la philosophie naturelle médiévale,* edited by Thomas Bénatouïl and Isabelle Draelants. Micrologus 40. SISMEL Edizioni del Galluzzo, 2011.

Sarton, George. *Introduction to the History of Science.* 3 vols. Johns Hopkins University Press, 1927–47.

———. "Review of *The Opus Majus,* by Roger Bacon, translated by Robert Belle Burke." *Isis* 11, no. 1 (1928): 138–41.

Schaffer, Simon. "Glass Works: Newton's Prisms and the Uses of Experiment." In *The Uses of Experiment: Studies in the Natural Sciences,* edited by David Gooding, Trevor Pinch, and Simon Schaffer. Cambridge University Press, 1989.

Schlund, P. Erhard. "Petrus Peregrinus von Maricourt: Sein Leben und seine Schriften; Ein Beitrag zur Roger-Bacon-Forschung." *Archivum Fransciscanum Historicum* 4 (1911): 436–55.

Schmitt, Jean-Claude. *The Holy Greyhound: Guinefort, Healer of Children Since the Thirteenth Century.* Translated by Martin Thom. Cambridge University Press, 1983.

Schwartz, Yossef. "Authority, Control, and Conflict in Thirteenth-Century Paris: Contextualizing the Talmud Trial." In *Jews and Christians in Thirteenth-Century France,* edited by Elisheva Baumgarten and Judah D. Galinsky. Palgrave Macmillan, 2015.

Seth, Suman. *Difference and Disease: Medicine, Race, and the Eighteenth-Century British Empire.* Cambridge University Press, 2018.

Shapin, Steven. "The House of Experiment in Seventeenth-Century England." *Isis* 79, no. 3 (1988): 373–404.

———. *The Scientific Revolution.* 2nd ed. University of Chicago Press, 2018.

Shapin, Steven, and Simon Schaffer. *Leviathan and the Air-Pump: Hobbes, Boyle, and the Experimental Life.* Princeton University Press, 1985.

Singer, Dorothea W. "Alchemical Writings Attributed to Roger Bacon." *Speculum* 7, no. 1 (1932): 80–86.

Smail, Daniel Lord. "In the Grip of Sacred History." *American Historical Review* 110, no. 5 (2005): 1337–61.

———. *On Deep History and the Brain.* University of California Press, 2008.

Smail, Daniel Lord, and Andrew Shryock. "History and the 'Pre.'" *American Historical Review* 118, no. 3 (2013): 709–37.

Smalley, Beryl. *The Study of the Bible in the Middle Ages.* 3rd ed. Blackwell, 1983 [1952].

Smith, A. Mark, ed. and trans. *Alhacen's Theory of Visual Perception: A Critical Edition, with English Translation and Commentary, of the First Three Books of Alhacen's 'De aspectibus,' the Medieval Latin Version of Ibn al-Haytham's Kitāb al-Manāẓir.* 2 vols. American Philosophical Society, 2001.

———. *From Sight to Light: The Passage from Ancient to Modern Optics.* University of Chicago Press, 2014.

Smith, Julian A. "Precursors to Peregrinus: The Early History of Magnetism and the Mariner's Compass in Europe." *Journal of Medieval History* 18, no. 1 (1992): 21–74.

Smith, Pamela H. *The Body of the Artisan: Art and Experience in the Scientific Revolution*. University of Chicago Press, 2004.

———. *From Lived Experience to the Written Word: Reconstructing Practical Knowledge in the Early Modern World*. University of Chicago Press, 2022.

Smoller, Laura Ackerman. *History, Prophecy, and the Stars: The Christian Astrology of Pierre d'Ailly, 1350–1420*. Princeton University Press, 1994.

Snyder, Laura J. "Discoverers' Induction." *Philosophy of Science* 64, no. 4 (1997): 580–604.

———. *Reforming Philosophy: A Victorian Debate on Science and Society*. University of Chicago Press, 2006.

———. "Renovating the *Novum Organum*: Bacon, Whewell, and Induction." *Studies in History and Philosophy of Science* 30, no. 4 (1999): 531–57.

———. "William Whewell." In *The Stanford Encyclopedia of Philosophy*, edited by Edward N. Zalta, Spring 2019. https://plato.stanford.edu/archives/spr2019/entries/whewell/.

Sohm, Philip. "Ordering History with Style: Giorgio Vasari on the Art of History." In *Antiquity and Its Interpreters*, edited by Alina Payne, Ann Kuttner, and Rebekah Smick. Cambridge University Press, 2000.

Sparavigna, Amelia Carolina. "Petrus Peregrinus of Maricourt and the Medieval Magnetism." *Mechanics, Materials Science, and Engineering* 12 (January 2016): 1–8.

Spiegel, Gabrielle M. "Épater les médiévistes." *History and Theory* 39, no. 2 (2000): 243–50.

———. *Romancing the Past: The Rise of Vernacular Prose Historiography in Thirteenth-Century France*. University of California Press, 1993.

Sprandel, Rolf. "World Historiography in the Late Middle Ages." Translated by Kristin E. Thomas. In *Historiography in the Middle Ages*, edited by Deborah Mauskopf Deliyannis. Brill, 2002.

Starn, Randolph. "The Early Modern Muddle." *Journal of Early Modern History* 6, no. 3 (2002): 296–307.

Steele, Robert. "Roger Bacon and the State of Science in the Thirteenth Century." In *Studies in the History and Method of Science*, edited by Charles Singer. Vol. 2. Clarendon Press, 1921.

———. "Roger Bacon as Professor: A Student's Notes." *Isis* 20, no. 1 (1933): 53–71.

Stephen, Leslie, ed. *Dictionary of National Biography*. Smith, Elder, and Co., 1885.

Stock, Brian. "The Self and Literary Experience in Late Antiquity and the Middle Ages." *New Literary History* 25, no. 4 (1994): 839–52.

Stoneman, Richard. *Alexander the Great: A Life in Legend*. Yale University Press, 2008.

Summers, David J. *Vision, Reflection, and Desire in Western Painting*. University of North Carolina Press, 2007.

Symes, Carol. "When We Talk About Modernity." *American Historical Review* 116, no. 3 (2011): 715–26.

Tachau, Katherine H. "God's Compass and *Vana Curiositas*: Scientific Study in the Old French *Bible Moralisée*." *Art Bulletin* 80, no. 1 (1998): 7–33.

———. "Some Aspects of the Notion of Intentional Existence at Paris, 1250–1320." In *Medieval Analyses in Language and Cognition*, edited by Stan Ebbesen and Russell L. Friedman. Royal Danish Academy of Sciences and Letters, 1999.

———. *Vision and Certitude in the Age of Ockham: Optics, Epistemology, and the Foundation of Semantics, 1250–1345*. Brill, 1988.

Teter, Magda. *Christian Supremacy: Reckoning with the Roots of Antisemitism and Racism.* Princeton University Press, 2023.

Thibodeaux, Jennifer D. *The Manly Priest: Clerical Celibacy, Masculinity, and Reform in England and Normandy, 1066–1300.* University of Pennsylvania Press, 2015.

Thiery, Antonio. "Federico II e la conoscenza scientifica." In *Intellectual Life at the Court of Frederick II Hohenstaufen*, edited by William Tronzo. National Gallery of Art, 1994.

Thomas of Kent. *Le Roman de Toute Chevalerie*, edited by Brian Foster and Ian Short. 2 vols. Anglo-Norman Text Society, 1976.

Thorndike, Lynn, Jr. *A History of Magic and Experimental Science.* 8 vols. Columbia University Press, 1923–58.

———. "Roger Bacon and Experimental Method in the Middle Ages." *Philosophical Review* 23, no. 3 (1914): 271–98.

Tierney, Brian. *The Crisis of Church and State: 1050–1300.* Medieval Academy Reprints for Teaching 21. University of Toronto Press, 1988.

Tipler, Paul A., and Ralph A. Llewellyn. *Modern Physics.* 6th ed. W. H. Freeman and Co., 2012.

Todhunter, Isaac, ed. *William Whewell, D. D., Master of Trinity College, Cambridge: An Account of His Writings with Selections from His Literary and Scientific Correspondence.* Macmillian, 1876.

Travaglia, Pinella. *Magic, Causality and Intentionality: The Doctrine of Rays in al-Kindi.* SISMEL Edizioni del Galluzzo, 1999.

Trevor-Roper, Hugh. *The Rise of Christian Europe.* Harcourt, Brace & World, 1965.

Trifogli, Cecilia. *Oxford Physics in the Thirteenth Century (ca. 1250–1270), Motion, Infinity, Place and Time.* Studien und Texte zur Geistesgeschichte des Mittelalters 72. Brill, 2000.

Truitt, E. R. "Celestial Divination and Arabic Science in Twelfth-Century England: The History of Gerbert of Aurillac's Talking Head." *Journal of the History of Ideas* 73, no. 2 (2012): 201–22.

———. "Knowledge and Power: Courtly Science and Political Utility in the Work of Roger Bacon." *Revista Española de Filosofía Medieval* 28, no. 1 (2021): 99–123.

———. "Making English Scientific: Chaucer, Translation, and the Astrolabe." *Isis* 115, no. 4 (2024): 757–75.

———. *Medieval Robots: Mechanism, Magic, Nature, and Art.* University of Pennsylvania Press, 2015.

———. "'Trei poëte, sages dotors, qui mout sorent di nigromance': Knowledge and Automata in Twelfth-Century French Literature." *Configurations* 12, no. 2 (2005): 167–93.

———. "The Virtues of Balm in Late Medieval Literature." *Early Science and Medicine* 14, no. 6 (2009): 711–36.

Twain, Mark [Samuel Langhorne Clemens]. *Life on the Mississippi.* James Osgood and Co., 1883.

Underwood, Cecil. "Roger Bacon." *Ave Maria*, June 20, 1914, 769–75.

Ungureanu, James C. *Science, Religion, and the Protestant Tradition: Retracing the Origins of Conflict.* University of Pittsburgh Press, 2019.

Van den Abeele, Baudoin. *La fauconnerie au Moyen Age: Connaissance, affaitage et médecine des oiseaux de chasse d'après les traités latins.* Klincksieck, 1994.

———. "Medieval Western and Eastern Treatises on Falconry in Comparison." In *Falconry in the Mediterranean Context During the Pre-Modern Era*, edited by Charles Burnett and Baudoin Van den Abeele. Droz, 2021.

van Dyke, Christina. "The Truth, the Whole Truth, and Nothing but the Truth: Robert Grosseteste on Universals (and the *Posterior Analytics*)." *Journal of the History of Philosophy* 48, no. 2 (2010): 153–70.

Van Nuffelen, Peter. *Orosius and the Rhetoric of History, Oxford Early Christian Studies.* Oxford University Press, 2012.

Vasari, Giorgio. *The Lives of the Artists.* Translated by Julia Conway Bondanella and Peter Bondanella. Oxford University Press, 1998.

———. *On Technique: Being the Introduction to the Three Arts of Design, Architecture, Sculpture and Painting.* Translated by Louisa S. Maclehose. Dover, 1960 [J. M. Dent, 1907].

Vattimo, Gianni. *The End of Modernity: Nihilism and Hermeneutics in Postmodern Culture.* Edited and translated by Jon R. Snyder. Johns Hopkins University Press, 1988.

———. *La fine della modernità.* Garzanti Editore, 1985.

Verger, Jacques. "The First French Universities and the Institutionalization of Learning: Faculties, Curricula, Degrees." In *Learning Institutionalized: Teaching in the Medieval University*, edited by John Van Engen. University of Notre Dame Press, 2000.

Vernon, Matthew X. *The Black Middle Ages: Race and the Construction of the Middle Ages.* Palgrave Macmillan, 2018.

Vescovini, Graziella Federici. "Alhazen vulgarisé: Le de li aspecti d'un manuscrit du Vatican (moitié du XIVe siècle) et le troisième Commentaire sur l'optique de Lorenzo Ghiberti." *Arabic Sciences and Philosophy* 8, no. 1 (1998): 67–96.

———. "La Fortune de l'optique d'Ibn Al-Haitham: *Le livre De aspectibus (Kitab al-manazir)* dans le moyen âge Latin." *Archives internationales d'histoire des sciences* 40 (1990): 220–38.

Warren, Michelle R. *Creole Medievalism: Colonial France and Joseph Bédier's Middle Ages.* University of Minnesota Press, 2011.

Weber, Max. "Science as Vocation." In *From Max Weber: Essays in Sociology*, edited and translated by H. H. Gerth and C. Wright Mills. Oxford University Press, 1946.

Wehr, Hans. *Dictionary of Modern Written Arabic.* Edited by J. Milton Cowan. Spoken Language Services, 1976.

Weill-Parot, Nicolas. "Encadrement ou dévoilement: L'occulte et le secret dans la nature chez Albert le Grand et Roger Bacon." *Micrologus: Natura, Scienze e Società Medievali* 14 (2006): 151–70.

———. "Knowability and the Power of *Ars*: Roger Bacon's Utopian Science as Anti-Magic." In *The Sciences of Roger Bacon's* Opus maius. Edited by Meagan S. Allen. Routledge, 2025.

Weinstock, Horst. "Roger Bacon und das 'hebräische Alphabet.'" *Aschkenas* 2 (1992): 15–48.

Westman, Robert S. *The Copernican Question: Prognostication, Skepticism, and Celestial Order.* University of California Press, 2011.

Whewell, William. *Architectural Notes on German Churches, with Remarks on the Origin of Gothic Architecture.* 2nd ed. J. & J. J. Deighton, 1835 [1830].

———. *An Elementary Treatise on Mechanics.* J. Smith . . . for J. Deighton & sons, 1819.

———. *History of the Inductive Sciences, from the Earliest to the Present Time.* 3 vols. 3rd ed. J. W. Parker, 1857 [1837].

———. "Letter to the Editor." *Edinburgh Review*, January 3, 1838.

———. *The Mechanics of Engineering: Intended for Use in Universities and in Colleges of Engineers.* John W. Parker, 1841.

White, Andrew Dickson. *A History of the Warfare of Science with Theology in Christendom.* 2 vols. D. Appleton and Co., 1896.

White, Hayden. *The Content of the Form: Narrative Discourse and Historical Representation.* Johns Hopkins University Press, 1987.

Whitney, Elspeth. "The Artes Mechanicae, Craftsmanship and the Moral Value of Technology." In *Design and Production in Medieval and Early Modern Europe: Essays in Honor of Bradford Blaine*, edited by Nancy van Deusen. The Institute of Mediaeval Music, 1998.

Wiedemann, Von Eilhard. "Arabische Studien über den Regenbogen." *Archiv für die Geschichte der Naturwissenschaften und der Technik* 4 (1912): 453–60.

———. "Zu Ibn al Haitams Optik." *Archiv für die Geschichte der Naturwissenschaften und der Technik* 3 (1912): 1–53.

Wiener, Carol Z. "The Beleaguered Isle: A Study of Elizabethan and Early Jacobean Anti-Catholicism." *Past and Present* 51 (1971): 27–62.

Wikander, Stig. "De l'Inde à l'Espagne: L'origine de la *Poridat de las Poridades*." In *Actas do IV Congresso de estudos árabes e islâmicos.* Brill, 1971.

William of Auvergne. *De universo.* In *Opera omni*, edited by Peter Aubouin. 2 vols. Paris, 1674.

Williams, Deanne. "*Friar Bacon and Friar Bungay* and the Rhetoric of Temporality." In *Reading the Medieval in Early Modern England*, edited by Gordon A. McMullan and David Matthews. Cambridge University Press, 2007.

Williams, Steven J. "Defining the *Corpus Aristotelicum*: Scholastic Awareness of Aristotelian Spuria in the High Middle Ages." *Journal of the Warburg and Courtauld Institutes* 58 (1995): 29–51.

———. "The Early Circulation of the Pseudo-Aristotelian *Secret of Secrets* in the West: The Papal and Imperial Courts." In *Le scienze alla corte di Federico II.* Micrologus 2. SISMEL Edizione del Galluzzo, 1994.

———. "Roger Bacon and His Edition of the Pseudo-Aristotelian *Secretum secretorum.*" *Speculum* 69, no. 1 (1994): 57–73.

———. "Roger Bacon in Context: Empiricism in the High Middle Ages." In *Expertus sum: L'expérience par les sens dans la philosophie naturelle médiévale*, edited by Thomas Bénatouïl and Isabelle Draelants. Micrologus 40. SISMEL Edizioni del Galluzzo, 2011.

———. *The Secret of Secrets: The Scholarly Career of a Pseudo-Aristotelian Text in the Latin Middle Ages.* University of Michigan Press, 2003.

———. "Two Independent Textual Traditions? The Pseudo-Aristotelian *Secret of Secrets* and the Alexander Legend." In *Trajectoires européennes du Secretum secretorum du Pseudo-Aristote (XIIIe-XVIe siècle)*, edited by Catherine Gaullier-Bougassas, Margaret Bridges, and Jean-Yves Tilliette. Brepols, 2015.

Wood, Anthony. *Survey of the Antiquities of the City of Oxford, Composed in 1661–6, by Anthony Wood*, edited by Andrew Clark. 3 vols. Oxford Historical Society, 1889–99.

Wood, Christopher S. *A History of Art History.* Princeton University Press, 2019.

Wood, Rega. "Roger Bacon: Richard Rufus' Successor as a Parisian Physics Professor." *Vivarium* 35, no. 2 (1997): 222–50.

Wootton, David. *Galileo: Watcher of the Skies.* Yale University Press, 2010.

———. *The Invention of Science: A New History of the Scientific Revolution.* HarperCollins, 2015.

Wrigglesworth, Edmund. "Roger Bacon." *Universe, an Illustrated Catholic Weekly Newspaper*, June 5, 1914.

Yates, Frances A. *Giordano Bruno and the Hermetic Tradition*. University of Chicago Press, 1964.

Yeo, Richard. *Defining Science: William Whewell, Natural Knowledge, and Public Debate in Early Victorian Britain*. Cambridge University Press, 1993.

Zambelli, Paolo. *The Speculum Astronomiae and Its Enigma: Astrology, Theology, and Science in Albertus Magnus and His Contemporaries*. Springer Dordrecht, 1992.

Index

Necessary Acknowledgments

This book has taken over a decade of research, writing, thinking, discussing, more research, rethinking, and rewriting to come into being. Institutions and individuals—scholars, archivists, librarians, students, mentors, family, and friends—have offered me expertise, help, advice, encouragement, critical feedback, and welcome distractions. It is a pleasure to recognize them and their efforts.

An initial short-term fellowship at The Huntington Library in San Marino, CA provided an opportunity to read early printed editions of Bacon's works, and I am grateful to the staff at the library for making them available to me. Likewise, I thank the staff and intellectual community at the Science History Institute, formerly the Chemical Heritage Foundation, in Philadelphia for their interest and excellent questions on my project when I first presented my early research findings there as a fellow. Carin Berkowitz, Elizabeth Berry-Drago, Michelle DiMeo, and Jim Voelkel helped create a welcoming and vibrant community; my fellow fellows Jason Pine and Meredith Farmer pointed me to reading in anthropology and Victorian science. This project developed its current contours during a stay at the Max-Planck-Institute for the History of Science in Berlin, first as a visiting fellow in Group II under the direction of Lorraine Daston, and was further refined during a productive series of conversations with members of the research group on premodern experience under the direction of Katja Krause. Thanks to Lorraine Daston for her comments and encouragement, and to Eric Gurevitch, Shireen Hamza, Florence Hsia, Ian Lawson, Elaine Leong, Omar Nasim, Nicola Polloni, Ahmed Ragab, Michael Stolberg, Matteo Valleriani, Christiane van Oetzen, and Dror Weil for their feedback. I was fortunate to spend a semester as a fellow in Fiesole at Villa I Tatti, the Harvard University Center for the Study of the Italian Renaissance, while reading and thinking about the invention of the Renaissance as a historical period, and my conversations with scholars there dramatically reshaped the book. I thank

Carmen Bambach, Yve-Alain Bois, Samir Boumedienne, Arianna Brunori, Anna Contadini, Veronica Copello, Angela Dressen, Mary Franklin-Brown, Paul Galvez, Dario Gamboni, Jane Hatter, Sundar Henny, Amanda Hilliam, Jocelyn Karman, Shawon Kinew, Dustin Klinger, Jakub Koguciuk, John Landsdowne, Lori di Lucia, Delia Moldovan, Cora Presezzi, Jonathan Regier, Clémence Revest, Michael Rocke, Éliane Roux, Joan-Pau Rubiés, Daniel Savoy, Jennifer Scappetone, Daniel Spaulding, Laura Stefanescu, Francesca Trivellato, Johanna Weis, and Guillermo Wilde, and the director, Alina Payne, for creating such a wonderful environment in which to work. Colleagues at the American Institute of Physics, the Bard Graduate Center, Brown University, Columbia University, Johns Hopkins University, Indiana University, University College London, the University of Maryland, the University of Virginia, and Yale University invited me to present my work in progress on this book, and I am grateful to Ahmed al-Rahim, Meagan S. Allen, Paola Bertucci, Adam Biggs, Alison Calhoun, Jonathan Conant, Hal Cook, Henry Cowles, Liam Daley, Mary Fissell, Yulia Frumer, Karl Galle, Shannon Gayk, Jeremy Greene, Patricia Ingham, Matt Jones, Fotini Kondyli, Irene Malfatto, Alex Medico More, Domenico Bertolini Meli, William Newman, Lawrence Principe, Joanna Radin, Eric Ramírez-Weaver, Bill Rankin, Kellie Robertson, Nancy Rogers, John Sabapathy, Matt Shindell, Pamela Smith, Matt Stanley, Will Thomas, Noèmie Tousaignant, Scott Truedell, John Warner, Ittai Weinryb, Simon Werrett, Josh White, and Michael Witmore for their intellectual generosity. Special thanks to Pamela Long for inviting me to participate in a workshop on machines in medieval and early modern Europe at Princeton University, and to Anthony Grafton, Samuel Holzman, Matthew C. Hunter, Jessica Keating, Craig Martin, Caroline Murphy, Steven Walton, Carolyn Yerkes, and Cristiano Zanetti for their feedback at a critical stage. Jeremiah Hackett and Yael Kedar invited me into the community of Bacon scholars for a special issue on Bacon in *Revista Española de Filosofía Medieval*. I also wish to thank the Andrew J. Mellon Foundation for awarding me a New Directions Fellowship to study Arabic and the medieval Islamicate world and I am grateful to Paul Cobb and Kaley Keener Zinaty at the University of Pennsylvania for welcoming me into their classrooms. *Elf shukur lihom.*

I could not have written this book without the help of numerous librarians, archivists, and manuscript curators: the staff at the library of the Royal Society; Trinity College, Cambridge; Magdalen College, Cambridge; Merton College, Oxford; the Weston Manuscript Library at the Bodleian Library,

Oxford; the British Library; Special Collections, Columbia University, and, of course, the Library of the University of Pennsylvania, including the wonderful staff in ILL and the Kislak Center for Special Collections. Thanks to Rupert Baker, Jonathan Bushell, Jonny Church, Evie Craggs, Ink David, Josephine Day, Mitch Fraas, Lauren Gala, Adam Green, Rebecca Hughes, Christopher Laico, Keith Moore, Hilary Moreton, Nick Okrent, John Pollack, Lynn Ransom, Diana Smith, and Lyle Waddell.

I've been enormously fortunate throughout my career to have learned from and worked alongside intellectually curious and thoughtful scholars. Katy Park, Joan Cadden, and the late John Murdoch helped me make sense of medieval natural philosophy; the late Mario Biagioli taught me to critique periodization and the idea of the "Scientific Revolution"; and Jan Ziolkowski encouraged my interest in medievalism. Ari Ariel, Assef Ashraf, Ignacio Gallup-Diaz, Anita Kurimay, Peter Magee, Kalala Ngalamulume, Sharon Ullman, and Alicia Walker, colleagues at Bryn Mawr College, helped shape this project through multiple conversations and chapter drafts, while Kim Cassidy and Mary Osirim helped me to secure necessary research funds. At the University of Pennsylvania, current and former colleagues Robby Aronowitz, David Barnes, Caz Batten, Etienne Benson, Shira Brisman, Melissa Charenko, Rita Copeland, Stephanie Dick, Ivan Drpić, Sebastián Gil-Riaño, Sarah Guérin, Zehra Hashmi, Nicholas Herman, Courtney Hicks, Rana Hogarth, Andi Johnson, John Kanbayashi, David Young Kim, Harun Küçük, Ada Kuskowski, Susan Lindee, Beth Linker, Amy Lutz, Jessica Martucci, Ramah McKay, Christine Mueller, Projit Mukharji, Kristen Ramsey, Emily Steiner, Beans Velocci, Heidi Voskuhl, and David Wallace helped me clarify my ideas and celebrated my progress. Thanks also to the students in my doctoral seminar on periodization and science: Zachary Endter, Sam Franz, Cheryl Hagan, Priya Nambrath, Valeria Seminario, and Qinyan Wu. Abigail Ballantyne, Anna Lehr Mueser, and Alex Tucker helped with research at critical junctures. My editor, Jenny Tan, believed in this project and helped me realize it; I thank her and her colleagues at Penn Press, especially Lily Palladino and Jonathan Repetti.

Other colleagues and friends far and wide played a big part in researching and writing this book. I am grateful to Jennifer Borland, Leah DeVun, Sonja Drimmer, William Eamon, Carissa Harris, Jack Hartnell, Darin Hayton, Timothy McCall, Colin Milburn, Sharrona Pearl, Jennifer Pruitt, and Eric Song for coming through with great advice and reading suggestions at critical moments. Mary Carruthers, Glenda Goodman, Colby Gordon,

Dave Kaiser, Michelle Karnes, Conrad Leyser, Henrietta Leyser, Julie Livingston, Aylin Malcolm, Pablo Maurette, James McHugh, Sam Roberts, Dominic Tierney, and Jessica Wolfe asked great questions; Claire Cairns, Heloisa von Tiesenhausen Cave, Thony Christie, Daniel Margocsy, Alex Marr, Natasha Mendelsohn, John Monks, Purnima Ramadorai, Tarun Ramadorai, Friederike von Tiesenhausen, Elizabeth Deans Walker, and Matthew Walker provided hospitality and welcome distractions on research trips. Extra special thanks to Aurelia Westlake, Clara Westlake, Kirsten Westlake, and Stian Westlake—my London family—for their open-hearted interest and support over the years. Tara Nummedal, Katharine Park, Alisha Rankin, Laura Stark, Jamie Taylor, John Tresch, and Nasser Zakariya read an early draft of this book in the darkest days of the pandemic and gave me generous feedback that helped me further clarify my ideas. Thank goodness for the solidarity of my writing group, The Randos: Danna Agmon, Elizabeth Popp Berman, Meredith Broussard, Jordan Ellenberg, Derek Gottlieb, Dan Hirschman, Christina Ho, Elizabeth Mazzolini, Siyanda Mohutsiwa, Amy Monroe, Deana Rohlinger, Tanya Schlam, Louise Seamster, Jennifer Sessions, Adam Slez, Rachel Swarns, Didem Türkoğlu, Jason Windawi, LaDale Winling, and Sarah Woulfin. Dale Booth, David Dorsey, Mike Droettboom, Josh Frost, Danielle George, Liz Greene, Josh Kletzkin, Phil Korshak, Mary Kate Leonard, Debby Levine, Jennifer McKee, Jennifer Paull, Blair Burns Potter, Bob Potter, Lillian Potter, Suzanne Schneider, Mat Schwartz, Kellie Staples, Steve Staples, Chris Timmerman, Marty Wasser, and Emily Yacus offered support, meals, sincere interest, and time with their pets and/or children.

My mother, Joan Greenbaum, and my extended network of siblings, niblings, aunts, uncles, and cousins have been so encouraging. Thank you to my mom for coming to a lot of online talks on Roger Bacon during the pandemic, and a few in person afterward, and for taking an interest in my work. Thank you to Kendall Barrett, Matthew Bretz, Sarah Bretz, Shawn Greenbaum, Randall Lancelot, Lindsey Truitt, Paul Truitt, Rebecca Truitt, and Sally Truitt; Jake Kantrowitz, Sarah Kantrowitz, Emily Maroni, Marsha Pinson, Max Pinson, Rachel Pinson, James Truitt, Stephen Truitt, Adam Vine, David Vine, and Joanne Vine; Caleb Barrett, Cheyenne Barrett, Grace Barrett, Randy Barrett, Tyson Barrett, Zack Barrett, Quintus Cunningham, Early Ferguson, Carson Garhart, Adelaide Greenbaum, Marlow Greenbaum, Thor Halvorson, Charlotte Hoffman, Lily Hoffman, Max Hoffman, Logan Howarth, Alexis Jesup, Alan Kirk, Brenna Lavelle,

Christy Halvorson Ross, Garth Ross, Jordan Sloat, Nat Stone, Antonia Stout, Alexander Truitt, and Rebecca Wodatch; David Berz, Andrea Grant, Jan McDavid, Gary McDavid, Sandy Merber, Jim Sneed, Marie Sneed, Cathy Weinberg, and David Weinberg; as well as Kerry Ahearn, Rob Dorit, Gillian Kendall, Callie Orszak, Tom Orszak, and Priscilla Southwell. Tick Ahearn offered encouragement and endless coffee, gave the gift of his attention, and uncomplainingly learned a great deal more about Roger Bacon than he bargained for.

Over the course of writing this book many dear family and friends whose interest and support mattered a great deal to me died. I remember Sherry Berz, June Dorsey, Parker Dorsey, Renée Dorsey, Griffin Ferguson, Bob Greenbaum, Sally Stone Halvorson, Will Noel, Jake Page, Susanne Stone Page, Tea Stiefel, Max O. Truitt, and John Vine. I wish they were alive to read these words.

Most fundamentally, this book was made possible by my access to affordable and high-quality medical care. While working on this project I had three major surgeries to ameliorate a life-threatening medical condition. I am beyond grateful to the people who helped me, before and after surgery, so that I could recover and write this book: Kelly Adorno, Joanne Bango, Keisha Cedeno, Aliyah Clay, Beth Ann Corr, Donna D'Angelo, Cathyann Huggins, Gyrid Lyon, Haajar Muhammad, Theresa Schweingruber, Kate Seethaler, and Yuhan Su. Most of all, I thank my neurosurgeons, M. Sean Grady and Jang Won Yoon. They and their dedicated colleagues—nurses, anesthesiologists, radiology technicians, occupational therapists, physiotherapists, medical assistants, housekeeping, and transport staff—provided excellent medical care without which I simply could not have completed this book.

www.ingramcontent.com/pod-product-compliance
Ingram Content Group UK Ltd.
Pitfield, Milton Keynes, MK11 3LW, UK
UKHW011658210826
474453UK00010B/51/J

9 781512 829884